Communications in Computer and Information Science   **769**

Commenced Publication in 2007
Founding and Former Series Editors:
Alfredo Cuzzocrea, Xiaoyong Du, Orhun Kara, Ting Liu, Dominik Ślęzak,
and Xiaokang Yang

Editorial Board

More information about this series at http://www.springer.com/series/7899

Mauro Dragoni · Monika Solanki
Eva Blomqvist (Eds.)

Semantic Web Challenges

4th SemWebEval Challenge at ESWC 2017
Portoroz, Slovenia, May 28 – June 1, 2017
Revised Selected Papers

 Springer

Editors
Mauro Dragoni
Fondazione Bruno Kessler
Povo, Trento
Italy

Eva Blomqvist
Linköping University
Linköping
Sweden

Monika Solanki
University of Oxford
Oxford
UK

ISSN 1865-0929 ISSN 1865-0937 (electronic)
Communications in Computer and Information Science
ISBN 978-3-319-69145-9 ISBN 978-3-319-69146-6 (eBook)
https://doi.org/10.1007/978-3-319-69146-6

Library of Congress Control Number: 2017956784

This Springer imprint is published by Springer Nature
The registered company is Springer International Publishing AG
The registered company address is: Gewerbestrasse 11, 6330 Cham, Switzerland

Preface

Common benchmarks, established evaluation procedures, comparable tasks, and public datasets are vital to ensure reproducible, evaluable, and comparable scientific results. To assess the current state of the art and foster the systematic comparison of contributions to the Semantic Web community, open challenges are now a key scientific element of the Semantic Web conferences. Following the success of the previous years, reflected by the high number of high-quality submissions, we organized the fourth edition of the "Semantic Web Challenges" as an official track of the ESWC 2017 conference (held in Portoroz, Slovenia, from May 28 to June 1, 2017), one of the most important international scientific events for the Semantic Web research community. The purpose of challenges is to validate the maturity of the state of the art in tasks common to the Semantic Web community and adjacent academic communities in a controlled setting of rigorous evaluation, thereby providing sound benchmarks, datasets, and evaluation approaches, which contribute to the advancement of the state of the art. This fourth edition included four challenges: Open Knowledge Extraction (OKE 2017), Semantic Sentiment Analysis (SSA 2017), Question Answering over Linked Data (QALD 7), and The Mighty Storage Challenge (MOCHA 2017). A total of 11 teams competed in the different challenges. The event attracted attendees from across the conference, with a high attendance for all challenge-related activities during ESWC 2017. This included the dedicated conference track and participation of challenge candidates during the ESWC poster and demo session. The very positive feedback and resonance suggests that the ESWC challenges provided a central contribution to the ESWC 2017 program.

This book includes the descriptions of all methods and tools that competed at Semantic Web Challenges 2017, together with a detailed description of the tasks, evaluation procedures, and datasets, offering to the community a snapshot of the advancement in those areas at that moment in time and material for replication of the results. The editors have divided the book content into four chapters, each dedicated to one area or challenge. Each chapter includes an introductory section by the challenge chairs providing a detailed description of the challenge tasks, the evaluation procedure, and associated datasets, and peer-reviewed descriptions of the participants' methods, tools, and results.

We would like to thank all challenge chairs, whose hard work during the organization of the 2017 edition of the Semantic Web Challenges helped in making it a fruitful event. Thanks to their work, we experienced a successful and inspiring scientific event, and we are now able to deliver this book to the community.

June 2017

Eva Blomqvist
Mauro Dragoni
Monika Solanki

Organization

Organizing Committee (ESWC)

General Chair

Eva Blomqvist   Linköping University, Sweden

Program Chairs

Diana Maynard   University of Sheffield, UK
Aldo Gangemi   Paris Nord University, France and ISTC-CNR, Italy

Workshops Chairs

Agnieszka Ławrynowicz   Poznan University of Technology, Poland
Fabio Ciravegna   University of Sheffield, UK

Tutorials Chairs

Anna Lisa Gentile   IBM Research Almaden, San Jose, CA, USA
Sebastian Rudolph   TU Dresden, Germany

Poster and Demo Chairs

Katja Hose   Aalborg University, Denmark
Heiko Paulheim   University of Mannheim, Germany

PhD Symposium Chairs

Rinke Hoekstra   Vrije Universiteit Amsterdam, The Netherlands
Pascal Hitzler   Wright State University, USA

Challenge Chairs

Monika Solanki   University of Oxford, UK
Mauro Dragoni   Fondazione Bruno Kessler, Italy

Semantic Technologies Coordinators

Lionel Medini   University of Lyon, France
Luigi Asprino   University of Bologna, Italy

EU Project Networking

Lyndon Nixon   Modul Universität Vienna, Austria
Maria Maleshkova   Karlsruhe Institute of Technology (KIT), Germany

Proceedings Chair

Olaf Hartig   Linköping University, Sweden

Process Improvement Chair

Derek Doran   Wright State University, USA

Publicity Chair

Ruben Verborgh   Ghent University, Belgium

Web Presence

Venislav Georgiev   STI, Austria

Treasurer

Alexander Wahler   STI, Austria

Local Organization

Marko Grobelnik   Jožef Stefan Institute, Slovenia

Challenges Organization

Challenge Chairs

Monika Solanki   University of Oxford, UK
Mauro Dragoni   Fondazione Bruno Kessler, Italy

Open Knowledge Extraction

René Speck   University of Leipzig, Germany
Michael Röder   University of Leipzig, Germany
Ricardo Usbeck   University of Leipzig, Germany
Axel-Cyrille Ngonga   Institute for Applied Informatics, Germany
  Ngomo
Horacio Saggion   Universitat Pompeu Fabra, Spain
Luis Espinosa-Anke   Universitat Pompeu Fabra, Spain
Sergio Oramas   Universitat Pompeu Fabra, Spain

Semantic Sentiment Analysis

Diego Reforgiato   University of Cagliari, Italy
  Recupero
Erik Cambria   Nanyang Technological University, Singapore
Emanuele Di Rosa   FINSA s.p.a, Italy

Question Answering over Linked Data

Ricardo Usbeck	University of Leipzig, Germany
Axel-Cyrille Ngonga Ngomo	Institute for Applied Informatics, Germany
Bastian Haarmann	Fraunhofer-Institute IAIS, Germany
Anastasia Krithara	National Center for Scientific Research Demokritos, Greece

Mighty Storage

Axel-Cyrille Ngonga Ngomo	Institute for Applied Informatics, Germany
Irini Fundulaki	Foundation for Research and Technology Hellas (FORTH), Greece
Mirko Spasic	OpenLink, UK
Henning Petzkam	Fraunhofer IAIS, Germany
Vassiliki Rentoumi	National Center for Scientific Research Demokritos, Greece

Contents

Semantic Sentiment Analysis

The Mighty Storage Challenge

MOCHA2017: The Mighty Storage Challenge at ESWC 2017

Kleanthi Georgala[1], Mirko Spasić[2(✉)], Milos Jovanovik[2], Henning Petzka[3], Michael Röder[1,4], and Axel-Cyrille Ngonga Ngomo[1,4]

[1] AKSW Research Group, University of Leipzig,
Augustusplatz 10, 04109 Leipzig, Germany
georgala@informatik.uni-leipzig.de
[2] OpenLink Software, London, UK
{mspasic,mjovanovik}@openlinksw.com
[3] Fraunhofer Institute IAIS, Schloss Birlinghoven, 53757 Sankt Augustin, Germany
henning.petzka@iais.fraunhofer.de
[4] DICE Group, Paderborn University, Pohlweg 51, 33098 Paderborn, Germany
{michael.roeder,axel.ngonga}@upb.de

Abstract. The aim of the Mighty Storage Challenge (MOCHA) at ESWC 2017 was to test the performance of solutions for SPARQL processing in aspects that are relevant for modern applications. These include ingesting data, answering queries on large datasets and serving as backend for applications driven by Linked Data. The challenge tested the systems against data derived from real applications and with realistic loads. An emphasis was put on dealing with data in form of streams or updates.

1 Introduction

Triple stores and similar solutions are the backbone of most applications based on Linked Data. Hence, devising systems that achieve an acceptable performance on real datasets and real loads is of central importance for the practical applicability of Semantic Web technologies. This need is emphasized further by the constant growth of the Linked Data Web in velocity and volume [1], which increases the need for storage solutions to ingest and store large streams of data, perform queries on this data efficiently and enable high performance in tasks such as interactive querying scenarios, the analysis of industry 4.0 data and faceted browsing through large-scale RDF datasets.

The lack of comparable results on the performance on storage solutions for the variety of tasks which demand time-efficient storage solutions was the main motivation behind this challenge. Our main aims while designing the challenge were to

- provide comparable performance scores for how well current systems perform on real tasks of industrial relevance and
- detect bottlenecks of existing systems to further their development towards practical usage.

© Springer International Publishing AG 2017
M. Dragoni et al. (Eds.): SemWebEval 2017, CCIS 769, pp. 3–15, 2017.
https://doi.org/10.1007/978-3-319-69146-6_1

2   The MOCHA Challenge

2.1   Overview

The MOCHA challenge was carried out within the Extended Semantic Web Conference (ESWC) 2017, which took place from May 28th, 2016 to June 1st, 2017 in Portoroz, Slovenia. Given the goals aforementioned, we designed the challenge to encompass the following tasks:

1. Ingestion of RDF data streams;
2. RDF data storage and
3. Browsing RDF data.

2.2   Tasks

Task 1: RDF Data Ingestion. The aim of task 1 was to measure the performance of SPARQL query processing systems when faced with streams of data from industrial machinery in terms of efficiency and completeness. Our benchmark ODIN (St**O**rage and **D**ata **I**nsertion be**N**chmark) was designed to test the abilities of tripe stores to store and retrieve streamed data. The experimental setup was hence as follows: We used a mimicking algorithm to generate RDF data similar to real data. We increased the size and velocity of RDF data used in our benchmarks to evaluate how well a given storage solution was able to deal with streamed RDF data derived from the mimicking approach aforemetionen. The data was generated from one or multiple resources in parallel and was inserted using SPARQL INSERT queries. SPARQL SELECT queries were used to check when the system completed the processing of the particular triples.

The input data for this task consists of data derived from mimicking algorithms trained on real industrial datasets. Each training dataset included RDF triples generated within a predefined period of time (e.g., a production cycle). Each event (e.g., each sensor measurement or tweet) had a timestamp that indicates when it was generated. The datasets differed in size regarding the number of triples per second. During the test, data was generated using data agents (in form of distributed threads). An agent is a data generator who is responsible for inserting its assigned set of triples into a triple store, using a SPARQL INSERT query. Each agent emulated a dataset that covered the duration of the benchmark. All agents operated in parallel and were independent of each other. As a result, the storage solution benchmarked had to support concurrent inserts. The insertion of a triple was based on its generation timestamp. To emulate the ingestion of streaming RDF triples produced within large time periods within a shorter time frame, we used a time dilatation factor that allowed rescaling data inserts to shorter timeframes. Our benchmark hence allows for testing the performance of the ingestion in terms of precision and recall by deploying datasets that vary in volume (size of triples and timestamps), and used different dilatation values, various number of agents and different size of update queries.

Task 2: Data Storage. The goal of task 2 was to measure how data storage solutions perform with interactive, simple, read, SPARQL queries as well as complex ones. We used our Data Storage Benchmark (DSB)[1] for this purpose. DSB runs simple and complex SPARQL SELECT queries, accompanied with a high insert data rate via SPARQL UPDATE queries, in order to mimic real use-cases where READ and WRITE operations are bundled together. The queries were designed to stress the system under test in different choke-point areas, while being credible and realistic.

The dataset used with DSB is an RDF dataset derived from the LDBC Social Network Benchmark (SNB) dataset[2], but modified so that its characteristics further match real-world RDF datasets [4]. These modifications were considered necessary due to the high structuredness of the SNB RDF dataset, which makes it more similar to RDB dataset, as opposed to real-world RDF datasets, such as DBpedia. Our DSB dataset was pre-generated in several different scale factors (sizes), and split in two parts: the dataset that should be loaded by the system under test, and a set of update streams containing update queries.

The benchmark was started by loading the dataset into the data storage solution under test, after which the benchmark queries were executed. The execution format was defined as a *query mix* which mimics the activities of a real-world online social network, e.g. there were more executions of short lookup queries than complex queries, each complex query was followed by one or more short lookups, etc. The current version of DSB supports sequential execution of the queries, which allows the storage system to use all available resources for the current query. As a final step, the results from the executed queries were evaluated against expected results. The main KPIs of this task were bulk loading time, average task execution time, average task execution time per query type, number of incorrect answers, and throughput.

Task 3: Faceted Browsing. The task on faceted browsing checked existing solutions for their capabilities of enabling faceted browsing through large-scale RDF datasets. Faceted browsing stands for a session-based (state-dependent) interactive method for query formulation over a multi-dimensional information space, where it is the efficient transition from one state to the next that determines the user's experience. The goal was to measure the performance relative to a number of types of transitions, and thereby analysing a system's efficiency in navigating through large datasets. We made use of the Benchmark on Faceted Browsing[3] on the HOBBIT platform[4] to carry out the testing of systems.

As the underlying dataset, a transport dataset of linked connections was used. The transport dataset was provided by a data generator PoDiGG[5] [5] containing train connections between stations on an artificially created map. For

---

[1] https://github.com/hobbit-project/DataStorageBenchmark.
[2] http://www.ldbcouncil.org/benchmarks/snb.
[3] https://github.com/hobbit-project/faceted-benchmark.
[4] https://master.project-hobbit.eu/.
[5] https://github.com/PoDiGG/podigg-lc.

the integration of delays into the dataset the Transport Disruption Ontology[6] [2] was used, which models possible events that can disrupt the schedule of transport plans. The dataset had to be loaded into the database of a participating system at the beginning of the benchmark.

A participating system was subsequently required to answer a sequence of SPARQL queries, which simulate browsing scenarios through the underlying dataset. The browsing scenarios were motivated by natural navigation behaviour of a user (such as a data scientist) through the data, as well as to check participating systems on a list of 14 choke points defined by certain types of transitions. The queries involved temporal (time slices), spatial (different map views) and structural (ontology related) aspects. In addition to these so-called instance retrievals, the benchmark included facet counts. Facet counts are SPARQL COUNT queries for retrieving the number of instances behind a certain facet selection, i.e. the number of instances that would remain after applying a certain additional filter restriction. This resulted in a total workload of 173 SPARQL queries divided up into 11 browsing scenarios.

3   Benchmarking Platform

All three tasks are carried out using the HOBBIT benchmarking platform[7]. This platform offers the execution of benchmarks to evaluate the performance of systems. For every task of the MOCHA challenge, a benchmark has been implemented. The benchmarks are sharing a common API which eases the work of the challenge participants.

For the benchmarking of the participant systems a server cluster has been used. Each of these systems could use up to three servers of this cluster each of them having 256 GB RAM and 32 cores. This enabled the benchmarking of monolythical as well as distributed solutions.

4   The Challenge

4.1   Overview

The MOCHA2017 challenge ran on 22nd May, 2017 and its results were presented during the ESWC 2017 closing ceremony. Three system participated in all tasks:

- *Virtuoso Open-Source Edition 7.2*[8], developed by OpenLink Software, that served as the baseline system for all MOCHA2017 tasks (*MOCHA Baseline*),
- *QUAD*[9], developed by Ontos, and
- *Virtuoso Commercial Edition 8.0 (beta)* (see Footnote 8), developed by Open-Link Software.

---

[6] https://transportdisruption.github.io/.

[7] HOBBIT project webpage: http://project-hobbit.eu/ HOBBIT benchmarking platform: https://master.project-hobbit.eu HOBBIT platform source code: https://github.com/hobbit-project/platform.

[8] https://virtuoso.openlinksw.com/.

[9] http://ontos.com/.

4.2   Results and Discussion

Task 1: RDF Data Ingestion

KPIs Our evaluation consists of three KPIs:

– Recall, Precision and F-Measure: The INSERT queries created by each data generator were sent into a triple store by bulk load. After a stream of INSERT queries was performed against the triple store, a SELECT query was conducted by the corresponding data generator. In Information Retrieval, Recall and Precision were used as relevance measurements and were defined in terms of retrieved results and relevant results for a single query. Recall is the fraction of relevant documents that were successfully retrieved and precision is the fraction of the retrieved documents that are relevant to a query. F-measure is the harmonic mean of Recall and Precision. For our set of experiments, the relevant results for each SELECT query were created prior to the system benchmarking by inserting and querying an instance of the Jena TDB storage solution.

Additionally, we computed:

$$Macro\text{--}Average\text{--}Precision = \frac{\sum_{i=1}^{\lambda} Precision_i}{\lambda} \tag{1}$$

$$Macro\text{--}Average\text{--}Recall = \frac{\sum_{i=1}^{\lambda} Recall_i}{\lambda} \tag{2}$$

where λ is the number of SELECT queries performed against the storage solution during the execution of the benchmark and Micro and Macro-Average Recall, Precision and F-measure of the whole benchmark. The aforementioned measurements $Precision_i$ and $Recall_i$ are the precision and recall of the i-th SELECT query. We also calculated $Macro\text{--}Average\text{--}F\text{--}measure$ as the harmonic mean of Eqs. 1 and 2.

$$Micro\text{--}Average\text{--}Precision = \frac{\sum_{i=1}^{\lambda} |\{relevant\ results_i\} \cap \{retrieved\ results_i\}|}{\sum_{i=1}^{\lambda} |\{retrieved\ results_i\}|} \tag{3}$$

$$Micro\text{--}Average\text{--}Recall = \frac{\sum_{i=1}^{\lambda} |\{relevant\ results_i\} \cap \{retrieved\ results_i\}|}{\sum_{i=1}^{\lambda} |\{relevant\ results_i\}|} \tag{4}$$

where the $\{relevant\ results_i\}$ and $\{retrieved\ results_i\}$ are the relevant and the retrieved results of the i-th SELECT query resp. We also calculated $Micro\text{--}Average\text{--}F\text{--}measure$ as the harmonic mean of Eqs. 3 and 4.

We have to mention that misclassifications between the expected and received results does not necessarily mean that the triple stores are prone to misclassify results or to have a bad performance, but that there are miss-matches for results sets between Jena TDB and the storage solution.

- Triples per second: at the end of each stream and once the corresponding SELECT query was performed against the system, we measured the triples per second as a fraction of the total number of triples that were inserted during that stream. This was divided by the total time needed for those triples to be inserted (begin point of SELECT query - begin point of the first INSERT query of the stream). We provided the maximum value of the triples per second of the whole benchmark. The maximum triples per second value was calculated as the triples per second value of the last stream with Recall value equal to 1.
- Average answer time: we reported the average answer delay between the time stamp that the SELECT query has been executed and the time stamp that the results are send to the evaluation storage. The first aforementioned time stamp was generated by the benchmark when the SELECT query was sent to the system and the second time stamp was generated by the platform when the results of the corresponding SELECT query were sent to the storage.

Experiment set-up. ODIN require a set of parameters to be executed, that are independent of the triple store. For MOCHA, all three systems were benchmarked using the same values. Each triple store was allowed to communicate with the HOBBIT [3] platform for at most 25 mins. The required parameters and their corresponding values for MOCHA are:

- **Duration of the benchmark:** It determines the time interval of the streamed data. Value for MOCHA2017 = 600,000 ms.
- **Name of mimicking algorithm output folder:** The relative path of the output dataset folder. Value for MOCHA2017 = *output_data/*.
- **Number of insert queries per stream:** This value is responsible for determining the number of INSERT SPARQL queries after which a SELECT query is performed. Value for MOCHA2017 = 100.
- **Population of generated data:** This value determines the number of events generated by a mimicking algorithm for one Data Generator. Value for MOCHA2017 = 10, 000.
- **Number of data generators - agents:** The number of indepedent Data Generators that send INSERT SPARQL queries to the triple store. Value for MOCHA2017 = 4.
- **Name of mimicking algorithm:** The name of the mimicking algorithm to be invoked to generate data. Value for MOCHA2017 = *TRANSPORT_DATA*.
- **Seed for mimicking algorithm:** The seed value for a mimicking algorithm. Value for MOCHA2017 = 100.
- **Number of task generators - agents:** The number of indepedent Task Generators that send SELECT SPARQL queries to the triple store. Value for MOCHA2017 = 1.

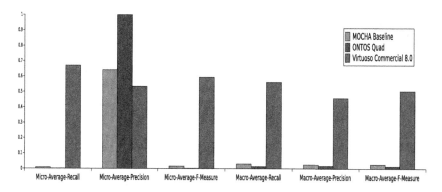

Fig. 1. Micro-Average-Recall, Micro-Average-Precision, Micro-Average-F-Measure, Macro-Average-Recall, Macro-Average-Precision, Macro-Average-F-Measure of *MOCHA Baseline, ONTOS Quad* and *Virtuoso Commercial 8.0* for MOCHA2017.

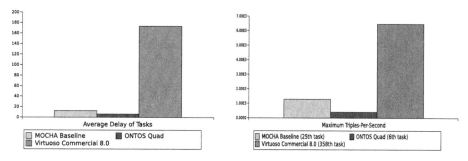

Fig. 2. Average Delay of tasks of *MOCHA Baseline, ONTOS Quad* and *Virtuoso Commercial 8.0* for MOCHA2017.

Fig. 3. Maximum Triples-per-Second of *MOCHA Baseline, ONTOS Quad* and *Virtuoso Commercial 8.0* for MOCHA2017.

Results for Task 1. By observing Fig. 1, we notice that *Virtuoso Commercial 8.0* has by far the best performance compared to the other two systems in terms of Macro and Micro-Average Precision, Recall and F-measure. *Virtuoso Commercial 8.0* was able to store and retrieve more triples through out the whole benchmark. However, the maximum performance value was achieved for Micro-Average Recall = 0.67, which indicates that the miss classifications between the Jena TDB and *Virtuoso Commercial 8.0* were still high on average. Additionally, since the Micro-Average values were higher compared to the Macro-Average values, we can conclude by stating that *Virtuoso Commercial 8.0* was able to retrieve more relevant triples to a SELECT query, for tasks with higher quantity of expected results.

Furthermore, we also notice that the Micro-Average Precision of *ONTOS Quad* is higher that the other systems. The Micro-Average values are calculated only when there are non-zero received results for a task. *ONTOS Quad* was able to retrieve results for the first 7 tasks and for the remaining 388 tasks, the system returned 0 received results or included 0 relevant results in its received result set, we notice that *Virtuoso Commercial 8.0* is the only system that was able to retrieve non-zero results for the majority of the SELECT queries.

In terms of maximum Triples-per-Second, based on Fig. 3, we notice that *Virtuoso Commercial 8.0* was able to achieve the highest maximum TPS at the latest task possible. It receives the last recall value of 1 at task 358 (out of 395), whereas the other systems have issues with recall at much earlier stages of the benchmark. Especially for the *ONTOS Quad* system, we see that its recall drops significantly after the 6th SELECT query.

Also, we need to mention that *ONTOS Quad* and *Virtuoso Commercial 8.0* were not able to perform all select queries within 25 mins. *ONTOS Quad* was not able to send results to the evaluation storage throughout the whole benchmark, whereas *Virtuoso Commercial 8.0* was not able to execute SELECT queries after 358 tasks, which is one of the reasons why its recall drops to 0.

Finally, we present the task delay for each task for all systems in Fig. 2. We notice that all systems have a relatively low task average delay over the set of SELECT queries. Whereas *Virtuoso Commercial 8.0* has a monotonically ascending task delay function, that drops to 0 after the 358th task, since the system is no longer available because it exceeded the maximum allowed time to process queries.

Task 2: Data Storage

KPIs. The main KPIs of this task are:

- **Bulk Loading Time:** The total time in milliseconds needed for the initial bulk loading of the dataset.
- **Average Task Execution Time:** The average SPARQL query execution time.
- **Average Task Execution Time Per Query Type:** The average SPARQL query execution time per query type.
- **Number of Incorrect Answers:** The number of SPARQL SELECT queries whose result set is different from the result set obtained from the triple store used as a gold standard.
- **Throughput:** The average number of tasks executed per second.

Experiment set-up. The Data Storage Benchmark has parameters which need to be set in order to execute the benchmark for this task. These parameters are independent of the triple store which is evaluated. The required parameters are:

- **Number of operations:** This parameter represents the total number of SPARQL queries that should be executed against the tested system. This number includes all query types: simple SELECT queries, complex SELECT queries and INSERT queries. The ratio between them, e.g. the number of queries per query type, has been specified in a query mix in such a way that each query type has the same impact on the overall score of the benchmark.

Fig. 4. Loading time   **Fig. 5.** Throughput

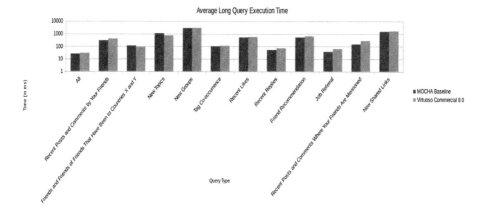

Fig. 6. Long queries

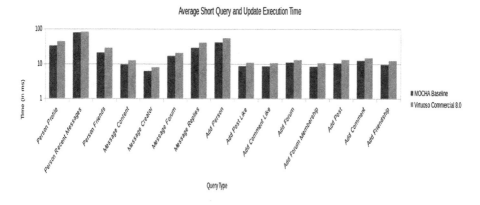

Fig. 7. Short queries and updates

This means that the simpler and faster queries are present much more frequently than the complex and slower ones. The value of this parameter for MOCHA was 15,000 operations.

– **Scale factor:** The DSB can be executed using different sizes of the dataset, i.e. with different scale factors. The scale factor for MOCHA was 1, i.e. the smallest DSB dataset.

Results for Task 2. Three systems applied and were submitted for Task 2: Virtuoso 7.2 Open-Source Edition by OpenLink Software, Virtuoso 8.0 Commercial Edition (beta release) by OpenLink Software, and QUAD by Ontos. Unfortunately, QUAD was not able to finish the experiment in the requested time (30 min), i.e. it exhibited a timeout.

Based on the results from the KPIs, shown in Figs. 4, 5, 6 and 7, the winning system for the task was Virtuoso 7.2 Open-Source Edition by OpenLink Software.

Task 3: Faceted Browsing

KPIs. For the evaluation, the received results from the participating system were compared with the expected ones. Results were returned in form of several key performance indicators:

The performance on instance retrievals was measured by a query-per-second score, by precision, recall and F1-score. Next to results for the full workload, the values were recorded for each of the 14 choke point individually. The list of choke points reads as follows:

1. Find all instances which (additional to satisfying all restrictions defined by the state within the browsing scenario) have a certain property value
2. Find all instances which (additionally) realize a certain property path with any value
3. Find all instances which (additionally) have a certain value at the end of a property path
4. Find all instances which (additionally) have a property value lying in a certain class
5. For a selected class that a property value should belong to, select a subclass
6. Find all instances that (additionally) have numerical data lying within a certain interval behind a directly related property
7. Similar to 6, but now the numerical data is indirectly related to the instances via a property path
8. Choke points 6 and 7 under the assumption that bounds have been chosen for more than one dimension of numerical data
9. Choke points 6,7,8 when intervals are unbounded and only an upper or lower bound is chosen
10. Go back to the instances of a previous step by unselecting previously chosen facets
11. Change the solution space to instances in a different class while keeping the current filter selections (Entity-type switch)
12. Choke points 3 and 4 with advanced property paths involved
13. Choke points 1 through 4 where the property path involves traversing edges in the inverse direction
14. Additional numerical data restrictions at the end of a property path where the property path involves traversing edges in the inverse direction

For facet counts, we measured the accuracy of participating systems in form of the deviation from the correct and expected count results. Additionally, we computed the query-per-second score for the corresponding COUNT queries.

Experiment set-up. The Faceted Browsing Benchmark required only one parameter which needed to be set in order to execute the benchmark for this task. This parameter consisted of a random seed, whose change alters the SPARQL queries of the browsing scenarios. The dataset was fixed and comprised about 1 million triples.

Results for Task 3. Three systems were submitted for Task 3: Virtuoso 7.2 Open-Source Edition by OpenLink Software which served as the MOCHA baseline, Virtuoso 8.0 Commercial Edition (beta release) by OpenLink Software, and QUAD by Ontos. Unfortunately, QUAD was not able to finish the experiment in the requested time (30 min), i.e. it exhibited a timeout. In Figs. 8 and 9, we display the results on instance retrievals. We see that both systems experienced problems on choke point number 12, which corresponds to filtering for the realisation of a certain property path (i.e., the task is to find all instances that, additionally to satisfying all restrictions defined by the state within the browsing scenario, realize a certain property path), and where the property path is of a rather complicated form. For example, complicated paths include those containing circles, or property paths where multiple entries need to be avoided.

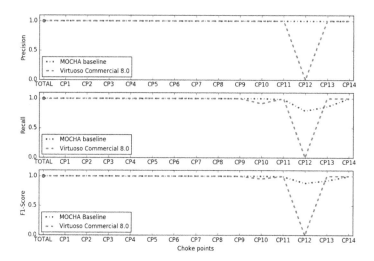

Fig. 8. Instance retrieval - accuracy

Consider now the query-per-second score in Fig. 9 for instance retrievals. Aside from the peculiar spike at choke point 2, the performance of both the open and the commercial version of Virtuoso are very similar with a slight advantage for the open source version. Interestingly, the query-per-second score of both system is the lowest for choke points 6 -8, which all correspond to selections of numerical data at the end of a property or property path.

In Figs. 10 and 11 we see the performance on count queries. Again, we see the slight advantage in the query-per-second score of the open source Virtuoso

Query-per-second score for each choke point

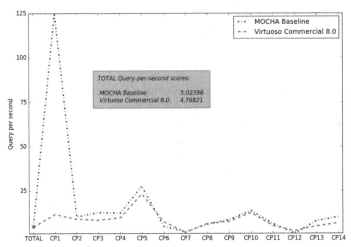

Fig. 9. Instance Retrieval - Query-per-second score

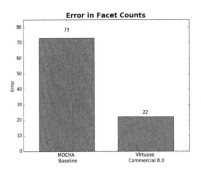

Fig. 10. Facet counts - accuracy

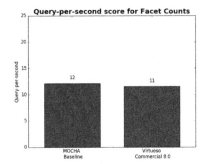

Fig. 11. Facet Counts - Query-per-second score

version serving as the MOCHA baseline. On the other hand, the commercial version of Virtuoso made less errors in returning the correct counts.

Overall, this resulted in a tie on this task, as both systems had very similar results with both having their slight advantage on one task or the other.

5　Conclusion

The goal of MOCHA2017 was to test the performace of storage solutions in terms of data ingestion, query answering and faceted browsing againist large datasets. We benchmarked and evaluated three triple stores and presented a detailed overview and analysis of our experimental set-up, KPIs and results. Overall, our results suggest that while the scalability of triple stores is improving, a need for scalable distributed solutions remains. As part of our future work, we will benchmark more triple storage solutions by scaling over the volume and velocity of the RDF data and use a diverse number of datasets to test the scalability of our approaches.

References

1. Auer, S., Lehmann, J., Ngonga Ngomo, A.-C.: Introduction to linked data and its lifecycle on the web. In: Polleres, A., d'Amato, C., Arenas, M., Handschuh, S., Kroner, P., Ossowski, S., Patel-Schneider, P. (eds.) Reasoning Web 2011. LNCS, vol. 6848, pp. 1–75. Springer, Heidelberg (2011). doi:10.1007/978-3-642-23032-5_1
2. Corsar, D., Markovic, M., Edwards, P., Nelson, J.D.: The transport disruption ontology. In: Arenas, M., et al. (eds.) ISWC 2015. LNCS, vol. 9367, pp. 329–336. Springer, Cham (2015). doi:10.1007/978-3-319-25010-6_22
3. Ngomo, A.-C.N., García-Rojas, A., Fundulaki, I.: HOBBIT: Holistic Benchmarking of Big Linked Data. ERCIM News, 2016(105) (2016)
4. Spasić, M., Jovanovik, M., Prat-Pérez, A.: An RDF dataset generator for the social network benchmark with real-world coherence. In: Proceedings of the Workshop on Benchmarking Linked Data (BLINK 2016), pp. 18–25 (2016)
5. Taelman, R., Verborgh, R., De Nies, T., Mannens, E.: PoDiGG: a public transport RDF dataset generator. In: Proceedings of the 26th International Conference Companion on World Wide Web, April 2017

Challenge Accepted:
QUAD Meets MOCHA2017

Alexander Potocki[✉], Daniel Hladky, and Martin Voigt

Ontos GmbH, Leipzig, Germany
{alexander.potocki,daniel.hladky,martin.voigt}@ontos.com

Abstract. Native RDF (http://www.w3.org/RDF/) stores have been making enormous progress in closing the performance gap compared to relational database management systems (RDBMS). But this small gap, however, still prevents the adoption of RDF stores in scenarios for large-scale enterprise applications. We solve this problem with our native RDF store QUAD and its fundamental design principles. It is based on a vector database schema for quadruples and it is realized by facilitating various index data structures. QUAD also comprises approaches to optimize the SPARQL query execution plan by using heuristic transformations. In this short paper, we briefly introduce QUAD and sketch in which tasks of the Mighty Storage Challenge we will attend to benchmark the current performance capabilities.

Keywords: RDF · SPARQL · Index · Query optimization · Benchmarking

1 Introduction

Over the past few years, we have seen explosive growth in the dissemination and use of semantic data. Initially, the traditional application fields of semantic technologies were areas as medicine, bioinformatics, public administration with their Linked Open Data portals. For the further establishment in other domains on enterprise-scale reliable and efficient solutions for storing and querying permanently increasing volumes of semantic data are the main foundation. Here, the *Mighty Storage Challenge* will contribute to a big extent.

Our goal is to provide an universal and customizable solution for storing semantic data that is efficient concerning its performance, does not require the use of a relational DB and translation of SPARQL into SQL. It needs to support recommendations of the World Wide Web Consortium (W3C) as RDF, SPARQL 1.1 [2], and SPARQL protocol[1]. Our RDF store **QUAD** [1] is the result of our ongoing research and development. In this introductory paper, we briefly describe our QUAD and explain how we are participating in the *Mighty Storage Challenge* to benchmark and compare the performance of our solution to other available stores.

---

[1] http://www.w3.org/standards/semanticweb/.

© Springer International Publishing AG 2017
M. Dragoni et al. (Eds.): SemWebEval 2017, CCIS 769, pp. 16–20, 2017.
https://doi.org/10.1007/978-3-319-69146-6_2

2   Related Works

Developing an RDF database, every developer faces some challenges. The two main problems are the following.

First, the choice, conceptualization, and development of a complete index set on SPO^2 triples or SPOG (see footnote 2) quadruples are complex. A series of works by Harth et al. [3–5], Baolin et al. [6], Weiss et al. in [7], and Abadi et al. [8] demonstrate similar methods of constructing them, using prefix search, reduce the full set of indices. Pursuing the elaboration of the multiple-index approach proposed in [3,9] and improved in the Hexastore solution [7], we have created a database structure supporting certain permutations of a set of elements for quads within SPOG (see footnote 2) relations.

The second challenge is the conceptualization and development of a query execution plan (QEP [14]). Several researchers formerly addressed the fundamental issues. Neumann et al. [10,11] proposed the query optimizer which mostly focuses on join order when generating execution plans and uses dynamic programming for the plan enumeration, with a cost model based on RDF-specific statistical synopses. Stocker et al. [12] presented SPARQL query optimization methods based on static optimization using the Basic Graph Pattern (BGP) method to optimize triple pattern join sequences. Gomathi et al. [13] described a multi-query optimization process that splits an input query into clusters using the K-means method based on the common sub-expression in the queries constituting an input set of queries. During our research, we examined equivalent query plan transformations based on heuristic rules worked out using computational complexities of algorithms for implementing operations, experimentally as well as through the observation of the system response times for various QEP configurations. This research direction was the most promising for us so that we have developed a set of heuristics never published in other works. We employ them for the static optimization of the original QEP.

3   QUAD: Design and Implementation

QUAD follows the generally accepted design of databases, which is sketched in Fig. 1. In order to receive and process queries from client applications, QUAD implements *SPARQL 1.1 Protocol*[3]. Before making a request to QUAD, any client application must authenticate itself using *Digest Access Authentication*[4] protocol. After the authentication process, QUAD creates a client session object and associates an access descriptor or, so called, authorization token with it, which determines the availability of particular data for subsequent client queries. Only then SPARQL queries are ready to be parsed and converted to the iterator tree in the *SPARQL Engine* module. The authorization token is accounted for by placing additional filter iterators in the tree. Leaf iterators are connected to

---

[2] Here, "S" stands for Subject, "P" for Predicate, "O" for Object, and "G" for Graph.

[3] https://www.w3.org/TR/sparql11-protocol/.

[4] https://en.wikipedia.org/wiki/Digest_access_authentication.

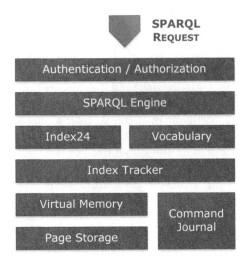

Fig. 1. Architectural design of QUAD.

a set of indexes that store different slices of the RDF data (*Index24*), as well as to the indexes of the literal values (*Vocabulary*). Indexes are implemented using the BTree algorithm. The nodes of the BTree are represented by blocks of memory or pages of given size specified during the configuration of the QUAD database. Each block has its unique numeric identifier. The *Virtual Memory* subsystem provides access to the pages by their identifiers and also caches them using the *2Q buffer cache algorithm* (Johnson et al. [15]). The *Page Storage* subsystem is responsible for loading and uploading data to permanent storage. This subsystem uses direct access to storage devices, bypassing the operating system's file cache to maximize performance. The *Index Tracker* tracks any changes to pages during the insertion, deletion or modification of data in BTree indexes. These changes are encoded by a set of incremental instructions, which in turn are stored on permanent storage by the *Command Journal* subsystem. These records, called the transaction log, can be used to restore the database in the event of an emergency shutdown.

As Fig. 1 illustrates, QUAD follows a component-based database design. Each component is described by its interface. The implementation details of the component are hidden from the other ones. Instances of components have unique identifiers. These identifiers serve to bind them to each other. The component life-cycle and their binding are managed by a specially developed framework that implements naming services, state storage services, and configuration services. QUAD is implemented in C++11 with intensive use of generic programming techniques. The architecture and operation system abstraction layer is performed mainly using the Boost library[5]. Assembly and testing were carried out on Linux

---

[5] http://www.boost.org/.

operating systems (Ubuntu and CentOS), Windows and Android, X86-64 and ARMv06 platforms.

4   Evaluation

To evaluate the performance of QUAD and compare it with other well-known RDF storages, we are going to participate in the *MIGHTY STORAGE CHALLENGE competition - ESWC 2017*[6]. This competition offers four types of tasks, for a comprehensive assessment of the performance of RDF storage. 1, 2 and 4 of these tasks are the general tests of intensive loading of RDF data in parallel mode and query execution over this data. These test scenarios emulate the database operating modes in real business tasks. The third task is related to the evaluation of the efficiency of storing versioned RDF data. QUAD does not contain any particular versioning implementations, so we can only emulate versioning using named graphs for different versions of RDF data. Since this approach is not efficient and may only offer a baseline performance, QUAD does not challenge this task.

For the competition, we prepared a special version of QUAD, configured and packed it into a docker image. The contest does not involve data durability testing, so we've disabled transaction journaling. For non-blocking data reads during the write operations, we activated the MVCC[7]. Almost all RAM is used for the index page cache. The number of threads executing simultaneous requests to the database corresponds to the number of processor cores in the system.

5   Conclusion and Further Work

In this introductive paper, we give a brief overview of our RDF store QUAD for the interested readers of the MOCHA2017 papers. If our native RDF store fits the challenge requirements, we are looking forward to the invitation to the tasks 1, 2 and 4 in order benchmark the already prepared dockerized version.

Besides the challenge, our ongoing work is to add features, stabilize them and boost the overall performance of QUAD. Regarding the latter, our primary focus is the development of a RDF data store cluster, which is geared towards the multi-platform processing of very-large-scale RDF datasets larger than 1 billions of triples. Therefore, we facilitate the concept of the parallel deployment of independent, full-featured RDF stores instance with a shared vocabulary index. Such an approach will prohibit the multiple storages of the same literal values in different stores, as well as to have a unique identification of RDF entities across all RDF stores in the cluster. One of the principal challenges in building a distributed database is QEP planning. Delays in transferring data between hosts can significantly reduce query performance. Hence, we developed a particular statistical index, which radically reduces the amount of data sent.

---

[6] https://project-hobbit.eu/challenges/mighty-storage-challenge.

[7] https://en.wikipedia.org/wiki/Multiversion_concurrency_control.

Acknowledgments. This work was partially supported by the BMWi project SAKE (Grant No. 01MD15006).

References

1. Potocki, A., Polukhin, A., Drobyazko, G., Hladky, D., Klintsov, V., Unbehauen, J.: OntoQuad: native high-speed RDF DBMS for semantic web. In: Klinov, P., Mouromtsev, D. (eds.) KESW 2013. CCIS, vol. 394, pp. 117–131. Springer, Heidelberg (2013). doi:10.1007/978-3-642-41360-5_10
2. Harris, S., Seaborne, A.: SPARQL 1.1 Query Language. Technical report, W3C Recommendation (2013). https://www.w3.org/TR/sparql11-query/
3. Harth, A., Decker, S.: Optimized index structures for querying RDF from the web. In: LA-WEB (Latin American Web Congress) (2005)
4. Harth, A., Umbrich, J., Hogan, A., Decker, S.: YARS2: a federated repository for querying graph structured data from the web. In: Aberer, K., et al. (eds.) ASWC/ISWC -2007. LNCS, vol. 4825, pp. 211–224. Springer, Heidelberg (2007). doi:10.1007/978-3-540-76298-0_16
5. Harth, A., Decker, S.: Yet Another RDF Store: Perfect Index Structures for Storing Semantic Web Data With Context, DERI Technical report (2004)
6. Baolin, L., Bo, H.: HPRD: a high performance RDF database. In: Li, K., Jesshope, C., Jin, H., Gaudiot, J.-L. (eds.) NPC 2007. LNCS, vol. 4672, pp. 364–374. Springer, Heidelberg (2007). doi:10.1007/978-3-540-74784-0_37
7. Weiss, C., Karras, P., Bernstein, A.: Sextuple Indexing for Semantic Web Data Management. PVLDB **1**(1), 1008–1019 (2008)
8. Abadi, D.J., Marcus, A., Madden, S., Hollenbach, K.J.: Scalable semantic web data management using vertical partitioning. In: VLDB, pp. 411–422 (2007)
9. Wood, D., Gearon, P., Adams, T.: Kowari: a platform for semantic web storage and analysis. In: XTeGh (2005)
10. Neumann, T., Weikum, G.: The RDF-3X engine for scalable management of RDF data. J. VLDB **19**(1), 91–113 (2010)
11. Neumann, T., Weikum, G.: RDF-3X: a RISC-style engine for RDF. PVLDB **1**(1), 647–659 (2008)
12. Stocker, M., Seaborne, A., Bernstein, A., Kiefer, C., Reynolds, D.: SPARQL basic graph pattern optimization using selectivity estimation. In: WWW 2008, pp. 595–604. ACM, New York (2008)
13. Gomathi, R., Sathya, C.: Efficient optimization of multiple SPARQL queries. IOSR J. Comput. Eng. (IOSR-JCE) **8**(6) (2013), pp. 97–101 (2013). www.iosr journals.org, e-ISSN: 2278–0661, p- ISSN: 2278–8727
14. Graefe, G.: Query evaluation techniques for large databases. ACM Comput. Surv. **25**(2), 73–170 (1993)
15. Johnson, T., Shasha, T.: 2Q: A low overhead high performance buffer management replacement algorithm. In: Proceedings of the 20th International Conference on Very Large Data Bases (VLDB 1994), San Francisco, CA, USA, pp. 439–450 (1994)

MOCHA 2017 as a Challenge for Virtuoso

Mirko Spasić[1,2(✉)] and Milos Jovanovik[1,3]

[1] OpenLink Software, London, UK
{mspasic,mjovanovik}@openlinksw.com
[2] Faculty of Mathematics, University of Belgrade, Belgrade, Serbia
[3] Faculty of Computer Science and Engineering,
Ss. Cyril and Methodius University in Skopje, Skopje, Macedonia

Abstract. The Mighty Storage Challenge (MOCHA) aims to test the performance of solutions for SPARQL processing, in several aspects relevant for modern Linked Data applications. Virtuoso, by OpenLink Software, is a modern enterprise-grade solution for data access, integration, and relational database management, which provides a scalable RDF Quad Store. In this paper, we present a short overview of Virtuoso with a focus on RDF triple storage and SPARQL query execution. Furthermore, we showcase the final results of the MOCHA 2017 challenge and its tasks, along with a comparison between the performance of our system and the other participating systems.

Keywords: Virtuoso · Social network benchmark · Mighty Storage Challenge · Benchmarks · Data storage · Linked Data · RDF · SPARQL

1 Introduction

Triple stores are the heart of a growing number of Linked Data applications. This uncovers a growing need for representative benchmarks which will fairly summarize their strengths and weaknesses [1], allowing stakeholders to choose between technologies from different vendors according to their needs and use-cases. The HOBBIT project[1] aims to push the development of Big Linked Data processing solutions by providing a family of industry-relevant benchmarks through a generic evaluation platform – the HOBBIT Platform [2]. In the scope of the project, several challenges are being organized, with the goal of reaching system providers, familiarizing them with the benchmarks of their interest, as well as the platform itself. The Mighty Storage Challenge (MOCHA 2017)[2] is one of these challenges: it aims to test the performance of systems capable of answering SPARQL SELECT queries and processing INSERT queries. Its goal is to provide objective measures for how well current systems perform on real tasks of industrial relevance and detect bottlenecks of existing systems to further their development towards practical usage. The challenge was accepted and presented

---

[1] https://project-hobbit.eu/.

[2] https://project-hobbit.eu/challenges/mighty-storage-challenge/.

© Springer International Publishing AG 2017
M. Dragoni et al. (Eds.): SemWebEval 2017, CCIS 769, pp. 21–32, 2017.
https://doi.org/10.1007/978-3-319-69146-6_3

in the Extended Semantic Web Conference (ESWC)[3] in 2017, held in Portoroz, Slovenia. Even though four tasks were initially planned, MOCHA 2017 consisted of three tasks in the end. OpenLink Software[4], with our RDF Quad Store – *Virtuoso 8.0 Commercial Edition (beta release)* – participated in all of three:

- Task 1: RDF Data Ingestion,
- Task 2: Data Storage, and
- Task 4: Browsing.

In Sect. 2, we will briefly present our system, Virtuoso, putting our focus on its quad storage, represented as a relational table, and its translation engine for converting SPARQL queries to SQL. We will describe all preparatory actions requested by the challenge organizers which the system had to fulfill in Sect. 3. After that, in Sect. 4, we will present the evaluation results of our system achieved during the challenge, along with a comparison with the outcomes of the other participants. Finally, Sect. 5 concludes the paper, and contains guidelines for future work and further improvement of our system.

2   Virtuoso Universal Server

Virtuoso[5] is a modern enterprise-grade solution for data access, integration, and relational database management. It is a database engine hybrid that combines the functionality of a traditional relational database management system (RDBMS), object-relational database (ORDBMS), virtual database, RDF, XML, free-text, web application server and file server functionality in a single system. It operates with SQL tables and/or RDF based property/predicate graphs. Virtuoso was initially developed as a row-wise transaction oriented RDBMS with SQL federation, i.e. as a multi-protocol server providing ODBC and JDBC access to relational data stored either within Virtuoso itself or any combination of external relational databases. Besides catering to SQL clients, Virtuoso has a built-in HTTP server providing a DAV repository, SOAP and WS* protocol end-points and dynamic web pages in a variety of scripting languages. It was subsequently re-targeted as an RDF graph store with built-in SPARQL and inference [3,4]. Recently, the product has been revised to take advantage of column-wise compressed storage and vectored execution [5].

The largest Virtuoso applications are in the RDF domain, with terabytes of RDF triples which do not fit into main memory. The excellent space efficiency of column-wise compression was the greatest incentive for the column store transition [5]. Additionally, this also makes Virtuoso an option for relational analytics. Finally, combining a schemaless data model with analytics performance is attractive for data integration in places with high schema volatility. Virtuoso has a shared cluster capability for scale-out. This is mostly used for large RDF deployments.

---

[3] https://2017.eswc-conferences.org/.
[4] https://www.openlinksw.com/.
[5] https://virtuoso.openlinksw.com/.

2.1   Triple Storage

The storage solution in Virtuoso is fairly conventional: a single table of four columns, named RDF_QUAD, holds one quad, i.e. a triple plus graph, per row. The columns are G for graph, P for predicate, S for subject and O for object. P, G and S are IRI IDs, for which Virtuoso has a custom data type, distinguishable at runtime from integer, even though internally this is a 32 or 64-bit integer. Since O is a primary key part, it is not desired to have long O values repeated in the index. Hence, Os of string type which are longer than 12 characters are assigned a unique ID and this ID is stored as the O of the quad table, while the mapping is stored in the RDF_IRI and RDF_PREFIX tables [3]. By default, and with the idea of faster execution, the table is represented as five covering indices, $PSOG$, $POSG$, SP, GS, and OP. In the first one, the quads are sorted primarily by predicate, then subject and object, and finally by graph. The structures of the other indices are analog to this one.

2.2   Compression

The compression is implemented at two levels. First, within each database page, Virtuoso stores distinct values only once and eliminates common prefixes of strings. Without key compression, there are 75 bytes per triple with a billion-triple LUBM[6] dataset (LUBM scale 8000). With compression, only 35 bytes per triple are present. Thus, when using 32-bit IRI IDs, key compression doubles the working set while sacrificing no random access performance. The benefits of compression are even better when using 64-bit IRI IDs [3].

The second stage of compression involves applying *gzip* to database pages, which reduces their size to a third, even after key compression. This is expected, since indices are repetitive by nature, even if the repeating parts are shortened by key compression [3].

2.3   Translation of SPARQL Queries to SQL

Internally, SPARQL queries are translated into SQL at the time of query parsing. If all triples are in one table, the translation is straightforward. In the next paragraph, we give a couple of simple SPARQL queries, and their simplified SQL translations.

All triple patterns from the SPARQL query should be translated to SQL as a self-join of the RDF_QUAD table, with conditions if there are common subjects, predicates and/or objects [3]. For example, if a SPARQL query asks for first and last names of 10 people, as shown in the example on Fig. 1, its SQL translation will be similar to the query given at Fig. 2. The functions __i2idn, __bft and __ro2sq are used for translation of RDF IRIs to the internal datatypes mentioned in the Subsect. 2.1, and vice versa.

A SPARQL *union* becomes an SQL *union* (Figs. 3 and 4) and *optional* becomes a left outer join (Figs. 5 and 6), while SPARQL *group by*, *having*,

---

[6] http://swat.cse.lehigh.edu/projects/lubm/.

```
 1 SELECT __ro2sq (t1.0) AS first, __ro2sq (t2.0) AS last
 2 FROM RDF_QUAD AS t0
 3     INNER JOIN RDF_QUAD AS t1
 4     ON ( t0.S = t1.S )
 5     INNER JOIN RDF_QUAD AS t2
 6     ON ( t0.S = t2.S AND t1.S = t2.S )
 7 WHERE
 8    t0.G = __i2idn ( __bft ( 'http://project-hobbit.eu/task2', 1))
 9    AND
10    t0.P = __i2idn ( __bft ( 'rdf:type', 1))
11    AND
12    t0.0 = __i2idn ( __bft ( 'snvoc:Person', 1))
13    AND
14    t1.G = __i2idn ( __bft ( 'http://project-hobbit.eu/task2', 1))
15    AND
16    t1.P = __i2idn ( __bft ( 'snvoc:firstName', 1))
17    AND
18    t2.G = __i2idn ( __bft ( 'http://project-hobbit.eu/task2', 1))
19    AND
20    t2.P = __i2idn ( __bft ( 'snvoc:lastName', 1))
21 LIMIT 10
```

```
1 SELECT ?first ?last
2 FROM <http://project-hobbit.eu/task2>
3 WHERE {
4    ?person a snvoc:Person .
5    ?person snvoc:firstName ?first .
6    ?person snvoc:lastName ?last .
7 }
8 LIMIT 10
```

Fig. 1. SPARQL query 1   **Fig. 2.** SQL translation of SPARQL query 1

```
 1 SELECT __ro2sq ( tmp.name ) AS name
 2 FROM (
 3     SELECT tmp1.0 AS name
 4     FROM RDF_QUAD AS tmp1
 5     WHERE
 6        tmp1.G = __i2idn ( __bft ( 'http://project-hobbit.eu/task2', 1))
 7        AND
 8        tmp1.P = __i2idn ( __bft ( 'snvoc:firstName', 1))
 9     LIMIT 10
10     UNION ALL
11     SELECT tmp2.0 AS name
12     FROM RDF_QUAD AS tmp2
13     WHERE
14        tmp2.G = __i2idn ( __bft ( 'http://project-hobbit.eu/task2', 1))
15        AND
16        tmp2.P = __i2idn ( __bft ( 'snvoc:lastName', 1))
17     LIMIT 10
18 ) AS tmp
19 LIMIT 10
```

```
1 SELECT ?name
2 FROM <http://project-hobbit.eu/task2>
3 WHERE {
4    { ?person snvoc:firstName ?name . }
5    UNION
6    { ?person snvoc:lastName ?name . }
7 }
8 LIMIT 10
```

Fig. 3. SPARQL query 2   **Fig. 4.** SQL translation of SPARQL query 2

order by and aggregate functions are translated to their SQL corresponding counterparts. Figure 4 shows an optimization trick: both members of a *union* have *limit* clauses, as well as the main *select*, providing that both parts of the query will not find more than 10 results.

In conclusion, a SPARQL query with n triple patterns will result with $n - 1$ self-joins. Thus, the correct join order and join type decisions are difficult to make given only the table and column cardinalities for the RDF triple or quad table. Histograms for ranges of P, G, O, and S are also not useful [3]. The solution is to go look at the data itself when compiling the query, i.e. do data sampling.

```
 1 SELECT __ro2sq ( tmp3.first ) AS first, __ro2sq ( tmp5.last ) AS last
 2 FROM RDF_QUAD AS tmp1
 3   LEFT OUTER JOIN (
 4     SELECT tmp2.S AS person,
 5            tmp2.O AS first
 6     FROM RDF_QUAD AS tmp2
 7     WHERE
 8       tmp2.G = __i2idn ( __bft ( 'http://project-hobbit.eu/task2', 1))
 9       AND
10       tmp2.P = __i2idn ( __bft ( 'snvoc:firstName', 1))
11   ) AS tmp3
12   ON ( tmp1.S = tmp3.person )
13   LEFT OUTER JOIN (
14     SELECT tmp4.S AS person,
15            tmp4.O AS last
16     FROM RDF_QUAD AS tmp4
17     WHERE
18       tmp4.G = __i2idn ( __bft ( 'http://project-hobbit.eu/task2', 1))
19       AND
20       tmp4.P = __i2idn ( __bft ( 'snvoc:lastName', 1))
21   ) AS tmp5
22   ON ( tmp1.S = tmp5.person )
23   WHERE
24     tmp1.G = __i2idn ( __bft ( 'http://project-hobbit.eu/task2', 1))
25     AND
26     tmp1.P = __i2idn ( __bft ( 'rdf:type', 1))
27     AND
28     tmp1.O = __i2idn ( __bft ( 'snvoc:Person', 1))
29 LIMIT 10
```

```
1 SELECT ?first ?last
2 FROM <http://project-hobbit.eu/task2>
3 WHERE {
4   ?person a snvoc:Person .
5   OPTIONAL { ?person snvoc:firstName ?first } .
6   OPTIONAL { ?person snvoc:lastName ?last } .
7 }
8 LIMIT 10
```

Fig. 5. SPARQL query 3   **Fig. 6.** SQL translation of SPARQL query 3

3   Challenge Prerequisites for Participation

In order to be a part of the challenge, the organizers proposed a set of requirements that participants had to conform to. The participants had to provide:

- A storage system that processes SPARQL INSERT queries
- A storage solution that can process SPARQL SELECT queries
- A solution as a Docker image that abides by the technical specifications, i.e.
  the MOCHA API

Virtuoso has build-in SPARQL support, so we only had to pack it as a Docker
image and develop a System Adapter, a component of the HOBBIT platform[7]
which implements the requested API and enables communication between the
benchmark and the Virtuoso instance. We developed an instance of the System Adapter for the commercial version of Virtuoso 8.0, which shares the same
Docker container with it. Its code is publicly available on GitHub[8].

After this component initializes itself, it starts receiving data from the Data
Generator, i.e. the files representing the benchmark dataset. When all files are
accepted, indicated by a signal from the Data Generator (the other part of
the platform that is in charge for creating the dataset for the benchmark), the
System Adapter starts loading the dataset into the Virtuoso instance. Upon
completion, it sends a signal to the other components indicating it is ready to
start answering the SPARQL queries, which are then sent by the Task Generator,

---

[7] http://master.project-hobbit.eu/.

[8] https://github.com/hobbit-project/DataStorageBenchmark.

a component which creates the tasks, i.e. the SELECT and INSERT queries. All accepted queries are then executed against our system, and their answers are sent to the Evaluation Storage, for validation against the expected answers and for measuring the achieved efficiency of the system.

4   Evaluation

In this section, we present the official results of the challenge for all its tasks.

4.1   Task 1: RDF Data Ingestion

The aim of this task is to measure the performance of SPARQL query processing systems when faced with streams of data from industrial machinery in terms of efficiency and completeness. This benchmark, called ODIN (St<u>O</u>rage and <u>D</u>ata <u>I</u>nsertion be<u>N</u>chmark), increases the size and velocity of RDF data used in order to evaluate how well can a system store streaming RDF data obtained from the industry. The data is generated from one or multiple resources in parallel and is inserted using SPARQL INSERT queries. At some points in time, the SPARQL SELECT queries check the triples that are actually inserted [6].

This task has three main KPIs:

- Triples per Second: For each stream, a fraction of the total number of triples that were inserted during that stream divided by the total time needed for those triples to be inserted.
- Average Answer Time: A delay between the time stamp that the SELECT query has been executed and the time stamp that the results are send to the Evaluation Storage.
- Correctness: A recall of each SELECT query by comparing the expected and retrieved results.

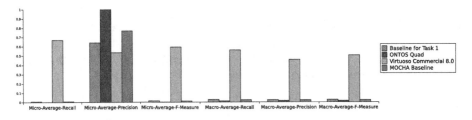

Fig. 7. Micro-Average-Recall, Micro-Average-Precision, Micro-Average-F-Measure, Macro-Average-Recall, Macro-Average-Precision, Macro-Average-F-Measure for Task 1 of MOCHA 2017.

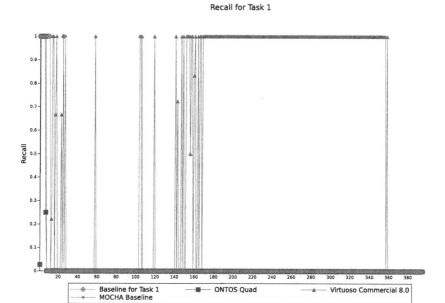

Fig. 8. Recall for task 1 of MOCHA 2017.

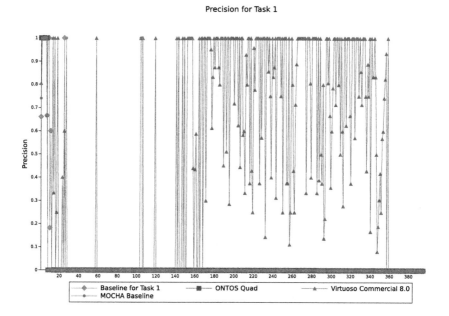

Fig. 9. Precision for task 1 of MOCHA 2017.

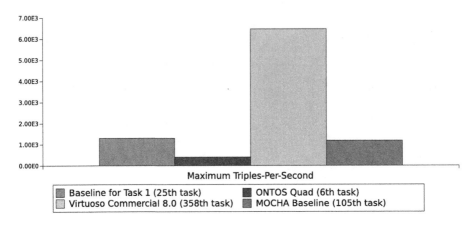

Fig. 10. Maximum triples-per-second for task 1 of MOCHA 2017.

Results: Here, we give the official results of our system, *Virtuoso 8.0 Commercial Edition (beta release)*, against ODIN, achieved during the challenge and published by the organizers of the challenge. The task organizers at MOCHA 2017 specified the benchmark parameters for the actual challenge run, in order to achieve the desired size and velocity of the RDF data. The values of the parameters were: number of insert queries per stream = 100, population of generated data = 10,000, number of data generators - agents = 4.

Our system, *Virtuoso Commercial 8.0*, had by far the best performance compared to the other systems, in terms of Macro and Micro-Average Precision, Recall, and F-measure (Fig. 7). By observing Figs. 8 and 9, it is obvious that our system was able to store and retrieve much more triples throughout the whole benchmark, than the other systems. In terms of maximum Triples-per-Second, based on Fig. 10, our system has just confirmed its convincing overall victory in this task, with a one order of magnitude better score. This results were announced by the organizers – our system had a best overall performance in terms of data ingestion and retrieval.

4.2 Task 2: Data Storage

The goal of this task is to measure how data storage solutions perform with interactive, simple, read, SPARQL queries as well as complex ones, accompanied with a high insert data rate via SPARQL UPDATE queries, in order to mimic real use-cases where READ and WRITE operations are bundled together. This task also tests systems for their bulk load capabilities [7].

The main KPIs of this task are:

- Bulk Loading Time: The total time in milliseconds needed for the initial bulk loading of the dataset.
- Throughput: The average number of tasks executed per second.

– Correctness: The number of SPARQL SELECT queries whose result set is different from the result set obtained from the triple store used as a gold standard.

Results: Based on the results from the KPIs, shown in Figs. 11, 12, 13 and 14, the winning system for the task was *Virtuoso 7.2 Open-Source Edition* by OpenLink Software, that was used as a baseline system for all tasks in the challenge. Our system, *Virtuoso 8.0 Commercial Edition*, was slightly slower, while the third system was not able to finish the experiment in the requested time, i.e. it exhibited a timeout, thus its scores are not present at the figures.

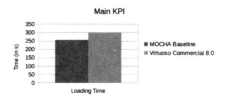

Fig. 11. Loading time for task 2 of MOCHA 2017.

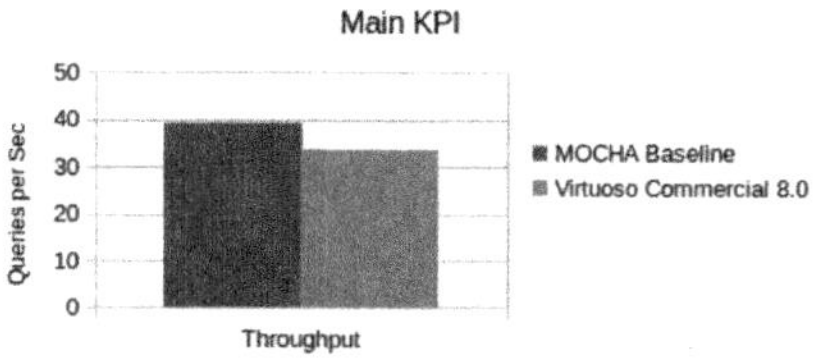

Fig. 12. Throughput for task 2 of MOCHA 2017.

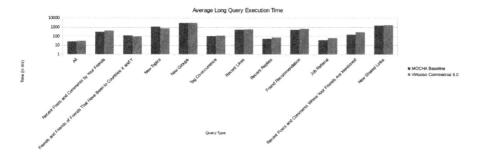

Fig. 13. Long queries for task 2 of MOCHA 2017.

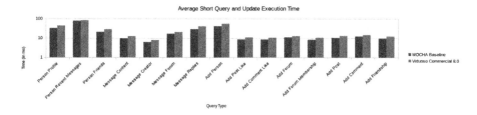

Fig. 14. Short queries and updates for task 2 of MOCHA 2017.

4.3   Task 4: Browsing

The task on faceted browsing checks existing solutions for their capabilities of enabling faceted browsing through large-scale RDF datasets, that is, it analyses their efficiency in navigating through large datasets, where the navigation is driven by intelligent iterative restrictions. The goal of the task is to measure the performance relative to dataset characteristics, such as overall size and graph characteristics [8].

The evaluation is based on the following performance KPIs:

– Throughput: The time required by the system is measured for the two tasks – facet count and instance retrieval – separately. The results are returned in a score function computing number of returned queries per second.
– Correctness: The facet counts are being checked for correctness. For each facet count, the distance of the returned count to the correct count in terms of absolute and relative value is recorded. For each instance retrieval the benchmark collects the true positives, the false positives and false negatives to compute an overall precision, recall and F1-score.

Results: Similar to the first task, the only two systems that managed to finish the task within the requested time slot are shown on the Figs. 15, 16, 17a and b, representing the main KPIs of the Faceted Browsing Benchmark. Based on that, the organizers announced a tie between our system and the baseline system. The Open-Source edition of Virtuoso was slightly faster, but the Commercial edition performed better on the correctness of the facet counts queries.

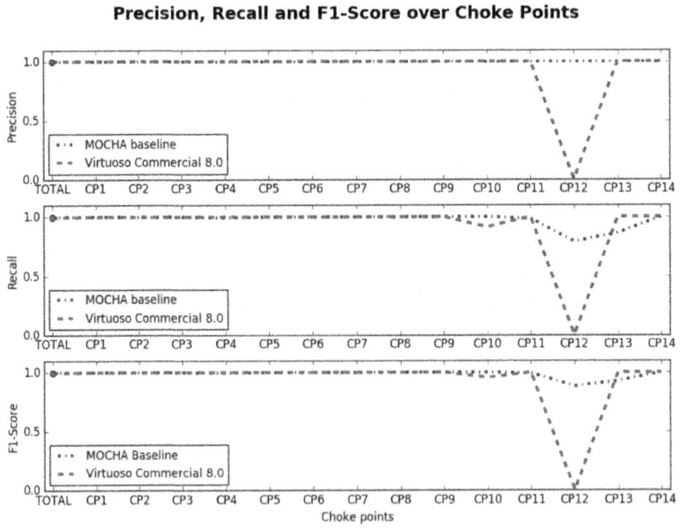

Fig. 15. Instance retrieval: correctness for task 4 of MOCHA 2017.

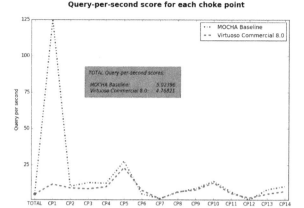

Fig. 16. Instance retrieval: speed for task 4 of MOCHA 2017.

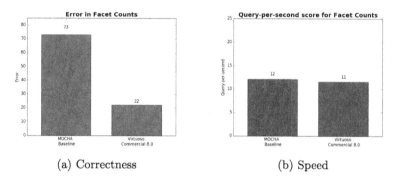

Fig. 17. Facet counts for task 4 of MOCHA 2017.

4.4 Overall Winner

As a summary, our system had a significant victory in Task 1; the baseline system was slightly better in Task 2; and there was a tie in Task 4. Based on the results in each task separately and the overall results, during the closing ceremony of the ESWC 2017 conference, the challenge organizers declared our system as the overall winner of the MOCHA 2017.

5   Conclusion and Future Work

This paper should be considered as an extended participant paper of MOCHA 2017, a challenge included in the Challenges Track of ESWC 2017, intended to test RDF storage and SPARQL systems for the following tasks: RDF Data Ingestion, Data Storage and Browsing. Thus, a short overview of Virtuoso Universal Server has been presented, with a focus on its RDF storage engine and the internal SPARQL to SQL translation. The evaluation part of the paper contains the official measurements from the challenge and its tasks. This section represents an excellent guideline as to where our Virtuoso optimizer should be improved.

As future work, a further evaluation has been planned against newer versions of the challenge benchmarks. For example, in Task 2, real-world workloads will be used, consisting of specified query mixes, where reads and updates are bundled together, and queries are run concurrently. Virtuoso will be tested with more dataset sizes and especially larger datasets, stressing its scalability. For this purpose, the HOBBIT platform will be used. We foresee improvements of the query optimizer, driven by the current evaluation.

Acknowledgments. This work has been supported by the H2020 project HOBBIT (GA no. 688227).

References

1. Morsey, M., Lehmann, J., Auer, S., Ngonga Ngomo, A.-C.: DBpedia SPARQL benchmark – performance assessment with real queries on real data. In: Aroyo, L., Welty, C., Alani, H., Taylor, J., Bernstein, A., Kagal, L., Noy, N., Blomqvist, E. (eds.) ISWC 2011. LNCS, vol. 7031, pp. 454–469. Springer, Heidelberg (2011). doi:10.1007/978-3-642-25073-6_29
2. Ngonga Ngomo, A.C., Röder, M.: HOBBIT: holistic benchmarking for big linked data. In: ERCIM News 2016 -105 (2016)
3. Erling, O., Mikhailov, I.: RDF support in the virtuoso DBMS. In: Pellegrini, T., et al. (eds.) Networked Knowledge - Networked Media: Integrating Knowledge Management, New Media Technologies and Semantic Systems 2009. SCI, vol. 221, pp. 7–24. Springer, Heidelberg (2009)
4. Erling, O., Mikhailov, I.: Virtuoso: RDF support in a native RDBMS. In: de Virgilio, R., et al. (eds.) Semantic Web Information Management: A Model-Based Perspective 2010, pp. 501–519. Springer, Heidelberg (2010)
5. Erling, O.: Virtuoso, a Hybrid RDBMS/Graph Column Store. https://virtuoso.openlinksw.com/dataspace/doc/dav/wiki/Main/VOSArticleVirtuosoAHybridRDBMSGraphColumnStore
6. Georgala, K.: Data Extraction Benchmark for Sensor Data. https://project-hobbit.eu/wp-content/uploads/2017/06/D3.1.1_First_Version_of_the_Data_Extraction_Benchmark_for_Sensor_Data.pdf
7. Jovanovik, M., Spasić, M.: First Version of the Data Storage Benchmark. https://project-hobbit.eu/wp-content/uploads/2017/06/D5.1.1_First_version_of_the_Data_Storage_Benchmark.pdf
8. Petzka, H.: First Version of the Faceted Browsing Benchmark. https://project-hobbit.eu/wp-content/uploads/2017/06/D6.2.1_First_Version_FacetedBrowsing.pdf

Open Knowledge Extraction Challenge

Open Knowledge Extraction Challenge 2017

René Speck[1(✉)], Michael Röder[1(✉)], Sergio Oramas[2(✉)],
Luis Espinosa-Anke[2(✉)], and Axel-Cyrille Ngonga Ngomo[1,3(✉)]

[1] AKSW Group, University of Leipzig, Leipzig, Germany
{speck,roeder}@informatik.uni-leipzig.de
[2] Universitat Pompeu Fabra, Barcelona, Spain
{sergio.oramas,luis.espinosa}@upf.edu
[3] University of Paderborn, Paderborn, Germany
ngonga@upb.de

Abstract. The Open Knowledge Extraction Challenge invites researchers and practitioners from academia as well as industry to compete to the aim of pushing further the state of the art of knowledge extraction from text for the Semantic Web. The challenge has the ambition to provide a reference framework for research in this field by redefining a number of tasks typically from information and knowledge extraction by taking into account Semantic Web requirements and has the goal to test the performance of knowledge extraction systems. This year, the challenge goes in the third round and consists of three tasks which include named entity identification, typing and disambiguation by linking to a knowledge base depending on the task. The challenge makes use of small gold standard datasets that consist of manually curated documents and large silver standard datasets that consist of automatically generated synthetic documents. The performance measure of a participating system is twofold base on (1) Precision, Recall, F1-measure and on (2) Precision, Recall, F1-measure with respect to the runtime of the system.

Keywords: Open Knowledge Extraction Challenge · Semantic Web

1 Introduction

The vision of the Semantic Web is an extension of the Document Web with the goal to allow intelligent agents a better reuse, sharing and understanding of the data in the Document Web. Agents are then able to automatically interpret the content of the Document Web. Thus, implementing the vision of the Semantic Web requires transforming unstructured and semi-structured data with knowledge extraction approaches from the Document Web into structured machine processable data for the current implementation of the Semantic Web, the Data Web.

In summary, we expect to trigger attention from the knowledge extraction community and foster their broader integration with the Semantic Web community with the Open Knowledge Extraction (OKE) challenge.

© Springer International Publishing AG 2017
M. Dragoni et al. (Eds.): SemWebEval 2017, CCIS 769, pp. 35–48, 2017.
https://doi.org/10.1007/978-3-319-69146-6_4

The rest of this paper is structured as follows: We begin with defining the OKE tasks in Sect. 2. In Sect. 3, we explain the evaluation with its datasets and scenarios. In Sect. 4 we give a brief introduction of the participating systems. In Sect. 5, we compare the results achieved by our evaluation on the gold and silver standard datasets. Finally, we discuss the insights provided by the challenge and possible extensions in Sect. 6.

2   Open Knowledge Extraction Challenge Tasks

The OKE challenge consist of three tasks. The first two tasks comprise named entity identification and named entity linking to the DBpedia knowledge base. For measuring the system performance in different perspectives based on the size and noise of the data, each of this two tasks is subdivided into two scenarios. The size of the data in scenario A is small and the data generation process was curated. In contrast, the size of the data in scenario B is large and the data generation process was carried out automatically with the help of BENGAL[1,2] to produce synthetic data.

The third task comprises named entity recognition and linking to Linked Brainz[3], the music knowledge base that is based on MusicBrainz[4]. This knowledge base is provided by the challenge (see Sect. 3.1) and dubbed MBL.

Both, the given input and the expected output are expressed with the help of the NIF[5][1] vocabularies and ontologies in an RDF serialisation. A participating system is not expected to process any preprocessing (e.g. pronoun resolution [2]) on the input data. In case a resource for an entity is missing in the knowledge base, a participating system is expected to generate a URI with the namespace of http://aksw.org/notInWiki/ for this emerging entity.

For carrying out the evaluation, this year the OKE challenge is using the HOBBIT benchmarking platform and the benchmark implementation of the HOBBIT project[6] which rely on the GERBIL evaluation framework [12].

2.1   Task 1: Focused Named Entity Identification and Linking

The first task aims at the identification and linking of entities of a given, limited set of entity types. It is a two-step process with the identification of named entities (Recognition) and the linking of those entities to resources in DBpedia (D2KB). A competing system is expected to identify named entity mentions in a given document by its start and end index, further to generate a URI to link each identified entity to DBpedia if possible or generate a URI for an emerging entity.

---

[1] http://github.com/aksw/bengal.
[2] http://project-hobbit.eu/wp-content/uploads/2017/04/D2.2.1.pdf.
[3] http://linkedbrainz.c4dmpresents.org/content/linkedbrainz-summary.
[4] http://musicbrainz.org.
[5] http://persistence.uni-leipzig.org/nlp2rdf.
[6] http://project-hobbit.eu.

The task is limited to a subset of resources in DBpedia, i.e., resources of the DBpedia ontology types: `Person`, `Place` and `Organisation`.

```
1  @prefix rdf: <http://www.w3.org/1999/02/22-rdf-syntax-ns#> .
2  @prefix xsd: <http://www.w3.org/2001/XMLSchema#> .
3  @prefix nif: <http://persistence.uni-leipzig.org/nlp2rdf/ontologies/
       nif-core#> .
4
5  <http://example.com/example-task1#char=0,91>
6    a nif:RFC5147String , nif:String , nif:Context ;
7    nif:beginIndex "0"^^xsd:nonNegativeInteger ;
8    nif:endIndex "124"^^xsd:nonNegativeInteger ;
9    nif:isString "Leibniz was born in Leipzig in 1646 and attended the
       University of Leipzig from 1661-1666."@en .
```
Listing 1.1. Example request document in task 1.

```
1  @prefix rdf: <http://www.w3.org/1999/02/22-rdf-syntax-ns#> .
2  @prefix xsd: <http://www.w3.org/2001/XMLSchema#> .
3  @prefix itsrdf: <http://www.w3.org/2005/11/its/rdf#> .
4  @prefix dbr: <http://dbpedia.org/resource/> .
5  @prefix nif: <http://persistence.uni-leipzig.org/nlp2rdf/ontologies/
       nif-core#> .
6
7  <http://example.com/example-task1#char=0,7>
8    a nif:RFC5147String , nif:String ;
9    nif:anchorOf "Leibniz"@en ;
10   nif:beginIndex "0"^^xsd:nonNegativeInteger ;
11   nif:endIndex "7"^^xsd:nonNegativeInteger ;
12   nif:referenceContext <http://example.com/example-task1#char=0,91> ;
13   itsrdf:taIdentRef dbr:Gottfried_Wilhelm_Leibniz .
14
15 <http://example.com/example-task1#char=20,27>
16   a nif:RFC5147String , nif:String ;
17   nif:anchorOf "Leipzig"@en ;
18   nif:beginIndex "20"^^xsd:nonNegativeInteger ;
19   nif:endIndex "27"^^xsd:nonNegativeInteger ;
20   nif:referenceContext <http://example.com/example-task1#char=0,91> ;
21   itsrdf:taIdentRef dbr:Leipzig .
22
23 <http://example.com/example-task1#char=53,74>
24   a nif:RFC5147String , nif:String ;
25   nif:anchorOf "University of Leipzig"@en ;
26   nif:beginIndex "53"^^xsd:nonNegativeInteger ;
27   nif:endIndex "74"^^xsd:nonNegativeInteger ;
28   nif:referenceContext <http://example.com/example-task1#char=0,91> ;
29   itsrdf:taIdentRef dbr:Leipzig_University .
```
Listing 1.2. Example of the expected response document in task 1.

Listing 1.1 is an example request document of task 1 and Listing 1.2 is the expected response document for the given request document. Both documents are formalized with NIF.

2.2 Task 2: Broader Named Entity Identification and Linking

This task extends the former task towards the DBpedia ontology types. Beside the three types of the first task, a competing system might have to identify other types of entities and to link these entities as well. In the first column in Table 1, a complete list of types that are considered in this task is provided. The middle column contains example subtypes of the corresponding class if any such class is available and the last column contains example instances in DBpedia for the related class respectively subtypes.

Table 1. Types, subtypes examples and instance examples for task 2.

Type	Subtypes	Instances
Activity	Game, Sport	Baseball,Chess
Agent	Organisation, Person	Leipzig_University
Award	Decoration, NobelPrize	Humanitas_Prize
Disease		Diabetes_mellitus
EthnicGroup		Javanese_people
Event	Competition, PersonalEvent	Battle_of_Leipzig
Language	ProgrammingLanguage	English_language
MeanOfTransportation	Aircraft, Train	Airbus_A300
PersonFunction	PoliticalFunction	PoliticalFunction
Place	Monument, WineRegion	Beaujolais, Leipzig
Species	Animal, Bacteria	Cat, Cucumibacter
Work	Artwork, Film	Actrius, Debian

2.3 Task 3: Focused Musical Named Entity Recognition and Linking

Task 3 composes of two subtasks (1) focused musical NE identification and classification and (2) linking to the MBL knowledge base that is based on MusicBrainz. Thus the domain of this task is music. A competing system has to fulfill both tasks in order to participate.

Listing 1.3 is an example input document and Listing 1.4 the expected annotated document for the given input, both formalized with NIF.

Task 3A: Focused Musical Named Entity Recognition. This subtask consists of the identification (Recognition) and classification (Typing) of named entities. The task is limited to a subset of resources in MBL, i.e., resources of the MBL ontology types: `Artist`, `Album` and `Song`. A competing system is expected to identify elements in a given text by its start and end index, further to assign one of the three types to each element.

Task 3B: Musical NE Linking. In this subtask a participating system has to link the recognised entities of the former subtask to the corresponding resources in MBL if existing or to generate a URI for the emerging entity.

```
1  @prefix rdf: <http://www.w3.org/1999/02/22-rdf-syntax-ns#> .
2  @prefix xsd: <http://www.w3.org/2001/XMLSchema#> .
3  @prefix nif: <http://persistence.uni-leipzig.org/nlp2rdf/ontologies/
       nif-core#> .
4
5  <http://example.com/example-task3#char=0,40>
6    a nif:RFC5147String , nif:String , nif:Context ;
7    nif:beginIndex "0"^^xsd:nonNegativeInteger ;
8    nif:endIndex "40"^^xsd:nonNegativeInteger ;
9    nif:isString "When Simon & Garfunkel split in 1970,..."@en .
```

Listing 1.3. Example request document in task 3

```
1  @prefix rdf: <http://www.w3.org/1999/02/22-rdf-syntax-ns#> .
2  @prefix xsd: <http://www.w3.org/2001/XMLSchema#> .
3  @prefix itsrdf: <http://www.w3.org/2005/11/its/rdf#> .
4  @prefix dbr: <http://dbpedia.org/resource/> .
5  @prefix mo: <http://purl.org/ontology/mo/> .
6  @prefix artist: <http://musicbrainz.org/artist> .
7  @prefix nif: <http://persistence.uni-leipzig.org/nlp2rdf/ontologies/
       nif-core#> .
8
9  <http://example.com/example-task3#char=5,22>
10   a nif:RFC5147String , nif:String ;
11   nif:anchorOf "Simon & Garfunkel"@en ;
12   nif:beginIndex "5"^^xsd:nonNegativeInteger ;
13   nif:endIndex "22"^^xsd:nonNegativeInteger ;
14   nif:referenceContext <http://example.com/example-task3#char=0,40> ;
15   itsrdf:taIdentRef artist:5d02f264-e225-41ff-83f7-d9b1f0b1874a ;
16   itsrdf:taClassRef mo:MusicArtist .
```

Listing 1.4. Example of the expected response document in task 3.

3   Evaluation

Overall, we follow two main evaluation approaches: subjective and objective. The subjective evaluation is based on paper reviews and the objective evaluation is based on computing relevance measures.

The knowledge bases DBpedia and MBL are used and the performance of a system is measured using Recall, Precision, F1-measure and β. Note that we reuse the ability of the GERBIL project enabling the benchmarking of systems that link to another knowledge base than DBpedia as long as there exist sameAs links between the two knowledge bases [8].

3.1 Datasets

The documents in the datasets might contain emerging entities, i.e., entities that are not part of the KB. These entities have to be marked and a URI has to be generated for them.

The datasets for the challenge are available at the challenge website[7]. Table 2 shows all the datasets available on the site assigned to the tasks and scenarios.

Table 2. Datasets.

Task	Scenario	File
1	A	Task1/A/training.tar.gz
		Task1/A/evaluation.tar.gz
	B	Task1/B/scenario-b-eval.zip
2	A	Task2/A/training.tar.gz
		Task2/A/evaluation.tar.gz
	B	Task2/B/scenario-b-eval.zip
3	A	Task3/A/training.tar.gz
		Task3/A/evaluation.tar.gz

The music knowledge base MBL used in task 3 is provided by the challenge at the website in the file MusicBrainzRDF.tar.gz as well.

3.2 Measures

Equations 1, 2, 3 and 4 formalize Precision p_d, Recall r_d, F1-measure f_d and beta β the performance measures we compute on the evaluation datasets for each document $d \in D$. They consist of the number of true positives TP_d, false positives FP_d and false negatives FN_d.

We micro average the performances over the documents[8].

$$p_d = \frac{TP_d}{TP_d + FP_d} \tag{1}$$

$$r_d = \frac{TP_d}{TP_d + FN_d} \tag{2}$$

---

[7] http://hobbitdata.informatik.uni-leipzig.de/oke2017-challenge/.

[8] The macro averages for the performance measures can be retrieved from the official HOBBIT SPARQL endpoint at http://db.project-hobbit.eu/sparql.

$$f_d = 2 \cdot \frac{p_d \cdot r_d}{p_d + r_d} \tag{3}$$

Let D be a set of documents for which β should be calculated. Let f_d be the F1-measure a benchmarked annotation system achieved for a given document $d \in D$ and let t_d be the time (in seconds) the annotation system needed for the annotation of d. Then the β value is the amount of F1-measure points a system achieves per second for a given amount of documents.

$$\beta = \frac{\sum_{d \in D} f_d}{\sum_{d \in D} t_d} \tag{4}$$

For matching the entity annotation positions of the benchmarked system and the correct entity markings of the datasets we used the *weak annotation matching* defined in [12]. Thus, an entity is counted as having the correct position, if its position overlaps with the correct position of the entity inside the dataset.

For example, our dataset considered "Franziska Barbara Ley". If a tool generated a URI for the emerging entity "Barbara Ley" and omitted "Franziska", it was assigned as a match.

3.3   Platform

The benchmark suite for named entity recognition and linking implemented within HOBBIT [4][9] reuses some of the concepts developed within the open-source project GERBIL. These concepts were migrated and adapted to the HOBBIT architecture. The Platform provides two different implementations of the benchmark described in the following subsections. It calculates values of Precision, Recall and F1-measure, measures the time a system needs to answer a request and counts the number of documents that cause errors in the benchmarked system.

Scenario A: Quality-Focused Benchmarking. The first type of benchmarking provided by our suite focuses on the measurement of quality a system achieves on a given set of documents. We assume that each benchmark dataset consists of a set of documents. The documents are sent to the benchmarked system one at a time. The benchmarked system generates a response and sends it back before receiving the next document. That means that the benchmarked system can be configured to concentrate all its resources on a single request and does not need to scale to a large number of requests. In this benchmarking, Scenario A, we rely on manually created gold standards.

The goal in this scenario is to achieve a high F1-measure in a quality-focused benchmarking.

---

[9] http://project-hobbit.eu/wp-content/uploads/2017/04/D2.2.1.pdf.

Scenario B: Performance-Focused Benchmarking. The second approach to benchmarking implemented by our platform aims to put a high load on the benchmarked system and to evaluate its runtime and quality in terms of Precision, Recall and F1-measure. This approach hence focuses on the ability of a system to annotate documents in parallel with an increasing amount of load.

The benchmark creates a large amount of synthetic documents from the given KB using BENGAL[10]. These documents are sent to the system in parallel without waiting for responses for previous requests but with predefined delays between the single documents. During a first phase of the benchmark, the generated work load equals 1 document per second. After the 80 documents of this first phase have been sent, the next phase is started using half of the delay of the previous time. This is done for 6 phases. In the seventh and last phase all 80 documents of the phase are sent without a delay, this leads to workloads of $\{1, 2, 4, 8, 16, 32, 80\}$ documents per second during the different phases.

The performance of a system is measured by β which is defined in Eq. 4. The scenarios goal is to achieve a high β value in a performance-focused benchmarking.

4 Participants

The challenge attracted four research groups. Two systems were not passing the subjective evaluation. The two remaining groups participated with there system in the challenge, ADEL and FOX.

4.1 Adel

ADEL [7], base on previous works [5,6], is an adaptive entity recognition and linking framework based on an hybrid approach that combines various extraction methods to improve the recognition level and an efficient knowledge base indexing process to increase the efficiency of the linking step. It deals with fine-grained entity types, either generic or domain specific. It also can flexibly disambiguate entities from different knowledge bases.

4.2 FOX

FOX [9] has been introduced in 2014 as an ensemble learning-based approach combining several diverse state of the art named entity recognition approaches and is based on the work in [3]. The FOX framework[11][10] outperforms the current state of the art entity recognizers. It relies on AGDISTIS [11] to perform named entity disambiguation. AGDISTIS is a pure entity linking approach (D2KB) based on string similarity measures, an expansion heuristic for labels to cope with co-referencing and the graph-based HITS algorithm. The authors published datasets[12] along with their source code and an API[13]. AGDISTIS can

---

[10] http://github.com/aksw/bengal.
[11] http://github.com/AKSW/FOX.
[12] http://github.com/AKSW/n3-collection.
[13] http://github.com/AKSW/AGDISTIS.

only be used for the D2KB task. Fox together with AGDISTIS can be use on the A2KB and the RT2KB task. Fox serves as the baseline system in this OKE challenge.

5   Results

In this section we present the results the participating systems reach on the three OKE challenge tasks. Tables 3 and 4 comprise the results for task 1 and 2 on both scenarios A and B. Tables 5 and 6 comprise the results for task 3A and 3B. The tables show the overall measures for Precision, Recall and F1-measure in the first three rows. The last two rows in each table show the averaged time in seconds a system needs to perform a document and the errors a system triggers. Further Tables 3, 4 and 5 show the interim results for step (i) in the next three rows and for step (ii) in the following three rows. For task 3.2 there are no interim results since there are no interim steps in this subtask.

5.1   Task 1

The measured values for scenario A in Table 3 show that ADEL outperforms Fox slightly with +1.09% F1-measure in step (i) Recognition. In step (ii) D2KB, Fox outperforms ADEL clearly with +16.82% F1-measure. Overall, Fox outperforms ADEL with +18.29% in Task 1 in scenario A.

In scenario B, the results are similar to scenario A. In step (i) ADEL outperforms Fox slightly as well as Fox outperforms ADEL clearly in step (ii).

Table 3. Results on task 1.

Experiment type	Micro measures	Scenario A		Scenario B	
		ADEL[a]	Fox[b]	ADEL[c]	Fox[d]
A2KB	Precision	33.24	53.61	18.28	59.12
	Recall	30.18	46.72	22.36	72.51
	F1-measure	31.64	**49.93**	20.12	**65.15**
Recognition	Precision	91.62	92.47	74.39	73.27
	Recall	83.20	80.58	90.98	89.85
	F1-measure	**87.21**	86.12	**81.85**	80.72
D2KB	Precision	40.15	61.96	28.03	93.87
	Recall	27.82	41.47	19.26	66.99
	F1-measure	32.87	**49.69**	22.83	**78.19**
	Time	7.98	6.98	231.31	179.29
	Errors	0	0	6	1

[a] http://w3id.org/hobbit/experiments#1497453653558.
[b] http://w3id.org/hobbit/experiments#1497440615203.
[c] http://w3id.org/hobbit/experiments#1497533785404.
[d] http://w3id.org/hobbit/experiments#1497533898908.

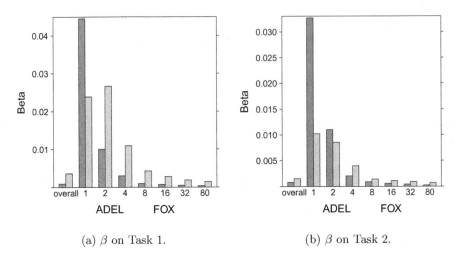

(a) β on Task 1.   (b) β on Task 2.

Fig. 1. β values on several numbers of requests and overall.

Overall, FOX reaches the highest value in scenario B with 65.15% F1-measure while ADEL reaches 20.12% F1-measure. With 6 and 1 errors, the error rates of ADEL and FOX are low compared to the number of 560 documents they had to annotate in this scenario.

Figure 1 depicts on the left side the detailed results for task 1 in scenario B. Surprisingly, ADEL reaches a clearly higher β value than FOX in the first phase for one document request per second. This is caused by the fast runtime of ADEL compensating its lower F1-score during that phase. In the following phases, the runtime of both systems increases—a clear sign that they are receiving requests to annotate document while they are still working on other documents. However, compared to FOX, the time that ADEL needs per document increases much more. Since the F1-score of both systems are similar over all phases but the time needed per document of FOX does not increase as much as it does for ADEL the β value of FOX remains higher than the value for ADEL. The observation of the increasing of processing time can be also seen in the comparison of the overall values of scenario A and B. While in A, ADEL needs 14% more time per document on average in scenario A this increases to 29% in scenario B. Together with the higher F1-score, the lower runtime of FOX leads to an overall β value which is four times higher than the value of ADEL.

5.2   Task 2

The measured values for scenario A in Table 4 show that ADEL outperforms FOX slightly with +4.83% F1-measure in step (i) Recognition. In step (ii) D2KB, FOX outperforms ADEL clearly with +14.02% F1-measure. Overall, FOX outperforms ADEL with +16.02% in Task 2 scenario A. In difference to Task 1, ADEL is nearly twice as fast as FOX in scenario A.

In scenario B, the results are similar to A. In step (i) ADEL outperforms FOX as well as FOX outperforms ADEL clearly in step (ii). Overall, FOX reaches the highest value in scenario B with 42.22% F1-measure while ADEL reaches 18.15% F1-measure.

Table 4. Results on task 2.

Experiment type	Micro measures	Scenario A		Scenario B	
		ADEL[a]	FOX[b]	ADEL[c]	FOX[d]
A2KB	Precision	31.40	56.15	17.44	44.90
	Recall	28.14	38.53	18.93	39.83
	F1-measure	29.68	**45.70**	18.15	**42.22**
Recognition	Precision	87.68	95.90	72.31	74.64
	Recall	78.57	65.80	78.50	66.21
	F1-measure	**82.88**	78.05	**75.27**	70.17
D2KB	Precision	39.93	63.42	28.57	82.38
	Recall	25.76	35.28	17.47	36.92
	F1-measure	31.32	**45.34**	21.68	**51.00**
	Time	4.60	7.66	261.48	245.99
	Errors	0	1	57	0

[a]http://w3id.org/hobbit/experiments#1497453720774.
[b]http://w3id.org/hobbit/experiments#1497440635319.
[c]http://w3id.org/hobbit/experiments#1497533810319.
[d]http://w3id.org/hobbit/experiments#1497533871062.

Figure 1 depicts on the right side the detailed results for task 2 for scenario B. Similar to task 1, ADEL reaches a clearly higher β value than FOX in the first two phases. This is again caused by the lower runtime of ADEL that compensates its lower F1-score. In all other phases FOX reaches a higher β value because as in Task 1 the runtime of ADEL increases much more than the runtime of FOX when it receives many requests in a short amount of time. Overall, FOX nearly reaches a β value twice as high as the value achieved by ADEL. It is also worth noting that this is the only experiment, in which the error rate of one of the systems is increased. For 57 of the 560 documents, ADEL responded with an error code. Nearly all of these errors—9, 26 and 21—occurred during the last three phases. Since the documents are chosen randomly and ADEL reported nearly no errors in the phases before, it is possible that they are related to the high load that ADEL receives during these phases.

5.3   Task 3

Task 3 is composed of two subtask, 3A and 3B. In the following, we first summarize the results on the first subtask and then on the second.

Table 5. Results on Task 3A.

Experiment type	Micro measures	ADEL[a]	FOX[b]
RT2KB	Precision	26.99	0
	Recall	27.24	0
	F1-measure	**27.12**	0
Recognition	Precision	35.03	63.02
	Recall	74.57	49.21
	F1-measure	47.66	**55.27**
Typing	Precision	64.33	0
	Recall	64.91	0
	F1-measure	**64.62**	0
	Time	37.19	7.82
	Errors	16	0

[a]http://w3id.org/hobbit/experiments#1497451343913.
[b]http://w3id.org/hobbit/experiments#1497449000101.

Table 6. Results on Task 3.2.

Experiment type	Micro measures	ADEL[a]	FOX[b]
D2KB	Precision	6.82	10.10
	Recall	5.10	4.97
	F1-measure	5.83	**6.66**
	Time	36.96	9.15
	Errors	16	0

[a]http://w3id.org/hobbit/experiments#1497453361862.
[b]http://w3id.org/hobbit/experiments#1497453423494.

Task 3A. The measured values for task 3A are depicted in Table 5. FOX reaches a higher F1-measure than ADEL, 55.27% to 47.66% in step (i). In step (ii) ADEL reaches a higher F1-measure, since FOX is not supporting this subtask due to the lack of the support of the music entity types.

Overall, ADEL reaches the highest value with 27.12% F1-measure on this task.

Task 3.2. The measured values for task 3.2 are depicted in Table 6. Both systems, ADEL and FOX, reach low performance on this task. ADEL achieves 5.83% and FOX a slightly higher value with 6.66%.

It is noteworthy that FOX processed the documents faster with 9.15s/doc in this subtask than ADEL with 39.96s/doc. Additionally, FOX encountered no errors in comparison to ADEL for which 16 errors have been reported.

5.4 Overall

The winner of Task 1 and 2 in both Scenarios A and B is Fox. For task 3A the winner is ADEL, since Fox is not supporting all subtasks. For task 3B the winner is Fox again. Since the advantage ADEL has in Task 3A is larger than the difference between Fox and ADEL in Task 3B, ADEL is the overall winner of Task 3.

The results on Task 1 and 2 suggest, that the Recognition component in ADEL achieved a higher F-measure than the respective component in Fox, but its linking component showed a worse performance than the respective component in Fox. Thus, it would be interesting to investigate the performance of the composition of the Recognition component of ADEL together with the linking component in Fox in this tasks.

The results on task 3 in the music domain suggest that the Recognition component of Fox achieved a better F-measure than ADEL. While Fox is not supporting the music entity types in its current version. Thus, it would be interesting to investigate the performance of an extended version that supports this types compared to ADEL in this task.

6 Conclusion

The Open Knowledge Extraction challenge attracted four research groups coming from Knowledge Extraction and Semantic Web communities. Indeed, the challenge proposal was aimed at attracting research groups from these two communities in order to further investigate exiting overlaps between Knowledge Extraction and the Semantic Web.

Although the participation in terms of the number of competing systems remained quite limited, we believe that the challenge is a breakthrough in the hybridisation of Semantic Web technologies with Knowledge Extraction methods. As a matter of fact, the evaluation framework is available online and can be reused by the community and for next editions of the challenge.

Acknowledgement. This work has been supported by the H2020 project HOBBIT (GA no. 688227) as well as the EuroStars projects DIESEL (project no. 01QE1512C) and QAMEL (project no. 01QE1549C). Also this work was partially funded by the Spanish Ministry of Economy and Competitiveness under the Maria de Maeztu Units of Excellence Programme (MDM-2015–0502).

References

1. Hellmann, S., Lehmann, J., Auer, S., Brümmer, M.: Integrating NLP using linked data. In: Alani, H., et al. (eds.) ISWC 2013. LNCS, vol. 8219, pp. 98–113. Springer, Heidelberg (2013). doi:10.1007/978-3-642-41338-4_7
2. Hobbs, J.: Pronoun resolution. Lingua **44**, 339–352 (1978)

3. Ngonga Ngomo, A.-C., Heino, N., Lyko, K., Speck, R., Kaltenböck, M.: SCMS – semantifying content management systems. In: Aroyo, L., Welty, C., Alani, H., Taylor, J., Bernstein, A., Kagal, L., Noy, N., Blomqvist, E. (eds.) ISWC 2011. LNCS, vol. 7032, pp. 189–204. Springer, Heidelberg (2011). doi:10.1007/978-3-642-25093-4_13

4. Ngonga Ngomo, A.-C., Röder, M.: HOBBIT: holistic benchmarking for big linked data. In: ESWC, EU Networking Session (2016)

5. Plu, J., Rizzo, G., Troncy, R.: A hybrid approach for entity recognition and linking. In: Gandon, F., Cabrio, E., Stankovic, M., Zimmermann, A. (eds.) SemWebEval 2015. CCIS, vol. 548, pp. 28–39. Springer, Cham (2015). doi:10.1007/978-3-319-25518-7_3

6. Plu, J., Rizzo, G., Troncy, R.: Enhancing entity linking by combining NER models. In: Sack, H., Dietze, S., Tordai, A., Lange, C. (eds.) SemWebEval 2016. CCIS, vol. 641, pp. 17–32. Springer, Cham (2016). doi:10.1007/978-3-319-46565-4_2

7. Plu, J., Troncy, R., Rizzo, G.: ADEL@OKE 2017: a generic method for indexing knowledge bases for entity linking. In: ESWC 2014, 14th European Semantic Web Conference, Open Extraction Challenge, 28th May-1st June 2017, Portoroz, Slovenia, May 2017

8. Röder, M., Usbeck, R., Ngonga Ngomo, A.-C.: Techreport for gerbil 1.2.2 - v1. Technical report, Leipzig University (2016)

9. Speck, R., Ngonga Ngomo, A.-C.: Ensemble learning for named entity recognition. In: Mika, P., Tudorache, T., Bernstein, A., Welty, C., Knoblock, C., Vrandečić, D., Groth, P., Noy, N., Janowicz, K., Goble, C. (eds.) ISWC 2014. LNCS, vol. 8796, pp. 519–534. Springer, Cham (2014). doi:10.1007/978-3-319-11964-9_33

10. Speck, R., Ngonga Ngomo, A.-C.: Named entity recognition using fox. In: International Semantic Web Conference 2014 (ISWC2014), Demos & Posters (2014)

11. Usbeck, R., Ngonga Ngomo, A.-C., Röder, M., Gerber, D., Coelho, S.A., Auer, S., Both, A.: AGDISTIS - graph-based disambiguation of named entities using linked data. In: Mika, P., Tudorache, T., Bernstein, A., Welty, C., Knoblock, C., Vrandečić, D., Groth, P., Noy, N., Janowicz, K., Goble, C. (eds.) ISWC 2014. LNCS, vol. 8796, pp. 457–471. Springer, Cham (2014). doi:10.1007/978-3-319-11964-9_29

12. Usbeck, R., Röder, M., Ngonga Ngomo, A.-C., Baron, C., Both, A., Brümmer, M., Ceccarelli, D., Cornolti, M., Cherix, D., Eickmann, B., Ferragina, P., Lemke, C., Moro, A., Navigli, R., Piccinno, F., Rizzo, G., Sack, H., Speck, R., Troncy, R., Waitelonis, J., Wesemann, L.: GERBIL - general entity annotation benchmark framework. In: 24th WWW Conference (2015)

ADEL@OKE 2017: A Generic Method for Indexing Knowledge Bases for Entity Linking

Julien Plu[1]([⊠]), Raphaël Troncy[1]([⊠]), and Giuseppe Rizzo[2]([⊠])

[1] EURECOM, Sophia Antipolis, France
{julien.plu,raphael.troncy}@eurecom.fr
[2] ISMB, Turin, Italy
giuseppe.rizzo@ismb.it

Abstract. In this paper we report the participation of ADEL to the OKE 2017 challenge. In particular, an adaptive entity recognition and linking framework that combines various extraction methods for improving the recognition level and implements an efficient knowledge base indexing process to increase the performance of the linking step. We detail how we deal with fine-grained entity types, either generic (e.g. Activity, Competition, Animal for Task 2) or domain specific (e.g. MusicArtist, SignalGroup, MusicalWork for Task 3). We also show how ADEL can flexibly link entities from different knowledge bases (DBpedia and MusicBrainz). We obtain promising results on the OKE 2017 challenge test dataset for the first three tasks.

Keywords: Entity recognition · Entity linking · Feature extraction · Indexing · OKE challenge · ADEL

1 Introduction

In this paper, we present our participation to the first three tasks of the OKE 2017 challenge, namely: (1) Focused NE Identification and Linking; (2) Broader NE Identification and Linking; (3) Focused Musical NE Recognition and Linking. The participation to these tasks has required to develop a system that can extract a broad range of entity types: generic in the Task 1, fine-grained in Task 2 or music-specific in Task 3. This has also triggered to develop a system that can handle multiple knowledge bases, such as DBpedia and MusicBrainz, to link the spotted candidates to referent resources.

We further develop the ADEL framework that is particularly suited to be adaptable to each of the requirements [3,4].

We improve the entity extraction and recognition process that includes a dictionary extractor that handles regular expressions.

We also propose a more sophisticated indexing process that allows to index the content of any RDF-based knowledge base such as DBpedia or Musicbrainz.

This paper mainly focuses on entity recognition and knowledge base indexing. Entity recognition refers to jointly performing the appropriate extraction

© Springer International Publishing AG 2017
M. Dragoni et al. (Eds.): SemWebEval 2017, CCIS 769, pp. 49–55, 2017.
https://doi.org/10.1007/978-3-319-69146-6_5

and typing of mentions. *Extraction* is the task of spotting mentions that can be entities in the text while *Typing* refers to the task of assigning them a proper type. *Linking* refers to the disambiguation of mentions in a targeted knowledge base. It is also often composed of two subtasks: generating candidates and ranking them accordingly to various scoring functions or link them to NIL if no candidates are found. Following the challenge requirements, we make use of the 2016-04 snapshot of DBpedia and a 2016-12 snapshot of Musicbrainz as the targeted knowledge bases.

The rest of the paper is structured as follows: Sect. 2 introduce our approach, Sect. 3 proposes the evaluations of this approach over each test dataset of OKE2017. Finally, conclusions and future work are discussed in Sect. 4.

2   Approach

In this section, we describe how we extract mentions from texts that are likely to be selected as entities by the *Extractor Module*. After having identified candidate mentions, we resolve their potential overlaps using the *Overlap Resolution Module*. Then, we describe how we disambiguate candidate entities coming from the extraction step. First, we create an index over the English DBpedia snapshot (version 2016-04) using the *Indexing Module*. This index is used to select possible candidates with the *Candidate Generation Module*. If no candidates are provided, this entity is passed to the *NIL Clustering Module*, while if candidates are retrieved, they are given to the *Linker Module*.

Extractor Module. We make use of five kinds of extractors: *(i)* Dictionary, *(ii)* POS Tagger, *(iii)* NER, *(iv)* Date, and *(v)* Number. Each of these extractors run in parallel. At this stage, an entity dictionary reinforces the extraction by bringing a robust spotting for well-known proper nouns or mentions that are too difficult to be extracted for the other extractors. We have developed a new approach for the dictionary extraction that consists in using a generic SPARQL query that retrieves all entity labels given a list of entity types. We developed a common API for these extractors based on Stanford CoreNLP [2] that is publicly available at https://github.com/jplu/stanfordNLPRESTAPI.

Indexing Module. An index can be seen as a two-dimensional array where each row is an entity in the index and each column is a property that describes the entity. Indexing the English DBpedia snapshot and retaining only properties that have literal values yields 281 columns. Once we have this index, we can search for a mention in this index and retrieve entity candidates. Searching, by default, over all columns (or properties used in the knowledge base), negatively impacts the performance of the index in terms of computing time. In order to optimize the index, we have developed a method that maximizes the coverage of the index while querying a minimum number of columns (or properties) [5]. For the DBpedia version 2016-04, there are exactly 281 properties that have literal values, while our optimization produced a reduced list of 8 properties: *dbo:wikiPageWikiLinkText, dbo:wikiPageRedirects, dbo:demonym,* *dbo:wikiPageDisambiguates*, *dbo:birthName*, *dbo:alias*, *dbo:abstract* and *rdfs:label*. This optimization drastically reduces the time of queryig by a factor of 4, in detail from 4 s to less than one second on a server that has 256 GB of RAM and a Intel Xeon CPU E5-2670 v3 @ 2.30 GHz. The source code of this optimization is also available[1]. Previously, we were using an index stored in Lucene. We have, however, observed unexpected behavior from Lucene such as not retrieving resources that partially match a query even if the number of results was not bound due to the lack of parameters and control of what can be searched on. The index is now built using *Elasticsearch* as a search engine that provides better scoring results. The indexing of a knowledge base follows a two-step process: *(i)* extracts the content of a knowledge base, and creates the Elasticsearch index; *(ii)* runs the optimization method in order to get the list of columns that will be used to query the index.

NIL Clustering Module. We propose to group the *NIL* entities (emerging entities) that may identify the same real-world thing. The role of this module is to attach the same *NIL* value within and across documents. For example, if we take two different documents that share the same emerging entity, this entity will be linked to the same *NIL* value. We can then imagine different *NIL* values, such as *NIL_1*, *NIL_2*, etc. We perform a string strict matching over each possible NIL entities (or between each token if it is a multiple token mention). For example, in sentence 23 of the datased used for Task 1, both the mention "Sully" and "Marine Jake Sully" will be linked to the same NIL entity.

Linker Module. This module implements an empirically assessed function that ranks all possible candidates given by the *Candidate Generation Module*:

$$r(l) = (a \cdot L(m, title) + b \cdot max(L(m, R)) + c \cdot max(L(m, D))) \cdot PR(l) \qquad (1)$$

The function $r(l)$ is using the Levenshtein distance L between the mention m and the title, and optionally, the maximum distance between the mention m and every element (title) in the set of Wikipedia redirect pages R and the maximum distance between the mention m and every element (title) in the set of Wikipedia disambiguation pages D, weighted by the PageRank PR, for every entity candidate l. The weights a, b and c are a convex combination that must satisfy: $a + b + c = 1$ and $a > b > c > 0$. We take the assumption that the string distance measure between a mention and a title is more important than the distance measure with a redirect page that is itself more important than the distance measure with a disambiguation page. In DBpedia not all pages have redirect or disambiguation pages associated, for this reason the two last elements of the formula are optional. This means that if a page does not have redirect pages, only the title and the disambiguation pages are evaluated, and the same logic is applied when only disambiguation pages exist, and finally, if no redirect and disambiguation pages exist, only the title is taken into account. In order to also apply this formula with Musicbrainz entities, we have computed a PageRank for each of them.

---

[1] https://gist.github.com/jplu/a16103f655115728cc9dcff1a3a57682.

3   Results and Discussion

In the OKE2107 challenge the evaluation for each task had two different scenarios: *(A)* the goal is to evaluate the performance of the linking by achieving the highest F1-score, *(B)* the goal is to evaluate the best ratio $\beta = \frac{F1-score}{runtime}$ of the system. The official OKE 2017 released scores for Scenario A are reported in Table 1. For Scenario B, the results are reported in Table 2. The comparison in each table is done with FOX [7,8] which was the baseline for each task. We have used the same ADEL configuration for each task:

1. Extraction: three different extractors, *(i)* Stanford NER with the 3-class, 4-class and 7-class models, *(ii)* Stanford NER with the model trained with the training set of the corresponding task, and *(iii)* a specific gazetteer made for the corresponding task.
2. Index: we use the DBpedia index for the two first tasks, and the Musicbrainz one for the third task. The Elasticsearch query we used to get the candidate has been adapted for each task.
3. Linking: we used the same weights for all tasks: $a = \frac{16}{21}$, $b = \frac{4}{21}$, and $c = \frac{1}{21}$.

This ultimately generated three different ADEL instances, one for each task.

Table 1. Results for scenario A over the OKE2017 datasets for the three tasks.

		ADEL			FOX		
		Precision	Recall	F1	Precision	Recall	F1
Task 1	Recognition	91.62	83.20	87.21	92.47	80.58	86.12
	D2KB	40.15	27.82	32.87	61.96	41.47	49.69
	A2KB	33.24	30.18	31.64	53.61	46.72	49.93
Task 2	Recognition	87.68	78.57	82.88	95.9	65.80	78.05
	D2KB	39.93	25.75	31.32	63.42	35.28	45.34
	A2KB	31.4	28.14	29.68	56.15	38.53	45.7
Task 3	Recognition	35.03	74.57	47.66	63.02	49.21	55.27
	Typing	64.33	64.91	64.62	0	0	0
	RT2KB	26.99	27.24	27.12	0	0	0

Concerning Task 1 and Task 2, the first thing we observe is the efficiency of the extraction part in ADEL, that of course can be leveraged depending the combination of extractors we use.

We can also observe that this extraction depends of what we want to extract, the more complex are the types to extract the more difficult is the extraction, and ADEL is robust against that, because despite the different number of entity types that must be extracted in Task 1 and Task 2, the F1 score shows a small difference between the two tasks and then proves ADEL robustness compared to FOX.

Table 2. Results for scenario B over the OKE2017 datasets for the two first tasks.

		ADEL			FOX		
		β	F1-score points	Avg millis per doc	β	F1-score points	Avg millis per doc
Task 1	Overall	0.0009	114.58	231314.48	0.0036	363.25	179287.18
	1	0.045	16.42	4613.25	0.024	50.17	26612.1
	2	0.01	16.49	20851.94	0.027	55	25808.69
	3	0.003	14.39	60331.46	0.011	55.43	63613.36
	4	0.00095	13.81	182078.99	0.0043	50.088	146824.36
	5	0.00074	19.65	337788.62	0.0028	53.63	240083.76
	6	0.00047	17.52	462000.34	0.0019	50.11	330158.19
	7	0.00038	16.3	567022.28	0.0015	48.82	420001.39
Task 2	Overall	0.00078	102.48	261497.17	0.0015	208.85	245985.25
	1	0.032	15.30	5849.78	0.01	29.28	35759.16
	2	0.011	14.9	17087.79	0.0085	26.71	39090.31
	3	0.002	15.26	93618.21	0.004	34.96	109790.31
	4	0.00088	18.17	258233.61	0.0014	26.72	237480.25
	5	0.00059	16.71	399959.27	0.0011	30.63	347780.25
	6	0.00041	12.06	538963	0.00092	31.76	433000.36
	7	0.00023	10.086	746879.68	0.0007	28.8	518996.09

Task 3 provides fine grained and specific entity types (artists, songs and albums), which bring a major issue: the name of an artist, a song or an album can be anything, including, for instance, a punctuation mark[2], for this reason we have preferred to configure ADEL to have a high recall.

For all tasks we observe a significant drop in performance at the linking stage. The linking formula is sensitive to the noise brought at the extraction step since this module does not take into account the entity context but instead relies on a combination of string distances and the PageRank global score. For example, in Task 1 dataset, sentence 1, the string distance score over the title, the redirect and the disambiguation pages between the mention *Trump* and the entity candidate `db:Trumpet` is higher than the correct entity candidate `db:Donald_Trump`.

We also evaluate the efficiency of our candidate generation module that, given a mention, should always provide the correct disambiguation link among a set of candidates. The evaluation is done as follows: from a training dataset, we perform a SPARQL query in order to get all mentions with their disambiguation link; then, for each mention, we query our index by using the list of columns listed in Sect. 2 to get a set of candidates and we check if the proper link is contained in that set. The minimum index of the correct link in this set is 1 while

---

[2] https://musicbrainz.org/work/25effd3c-aada-44d1-bcbf-ede30ef34cc0.

the maximum index is 1729 for Task 1, 1943 for Task 2, and 673 for Task 3. For Task 1 the recall@1729 is 94.65%, for Task 2 the recall@1943 is 90.22%, and for Task 3 the recall@673 is 97.32%. Most often, when the correct link is not retrieved, it is because the mention does not appear in the content of the queried columns, such as *007's*[3] in the sentence 37 of Task 1 dataset.

Regarding Scenario B in Table 2, we can see that ADEL has a drop of performance in terms of average millis per document from the 4th phase. In order to understand why this drop, we have profiled ADEL to detect the possible bottlenecks using the test dataset of Task 1. All the identified bottlenecks here are mostly observations that affect the runtime performance of ADEL. We succeeded to identify two significant bottlenecks: *(1)* the network latency, and *(2)* the candidate generation. The first is due to a high usage of external systems via HTTP queries (all the extractors and Elasticsearch), the sum of the latency of each HTTP query penalizes the runtime of ADEL. Unfortunately, we cannot really do something to solve this as it is an ADEL requirement to use external systems. Finally, the second bottleneck is Elasticsearch, arriving to a certain number of queries our ADEL instance gets stuck and starts to queue the queries. To solve this problem, we have developed a new architecture for our cluster by making each node able to be queried via a load balancer system. This solution allows to increase the number of queries run in the same time without being queued. This new architecture has divided the time to get our candidates by almost three (approximately one division per node).

4   Conclusion and Future Work

We have presented an entity extraction and linking framework that can be adapted to the entity types that have to be extracted and adapted to the knowledge base used to link the spotted entities. We have applied this framework to 3 tasks of the OKE 2017 challenge. While both recognition and the candidate generation processes provide good performance, the linking step is currently the main bottleneck in our approach. The performance drops significantly at this stage mainly due to a fully unsupervised approach.

We plan to investigate a new method that would modify Deep Structured Semantic Models [1] to make it compliant with knowledge bases and use it as a relatedness score between each candidate to build a graph composed of these candidates where each edge is weighted by this score. The path that has the highest score is chosen as the good one to disambiguate each extracted entity. This method should be agnostic to any knowledge base as it will use the relations among the entities. We also plan to align the entity types from different NER models, exploiting and extending previous work [6], in order to have a more robust recognition step. The association of multiples types of extraction techniques makes our approach extracting a significant amount of false positives. For this reason, we are also investigating to add a pruning step at the end of the process in order to reduce the amount of false positives. Finally, to improve

---

[3] http://dbpedia.org/resource/James_Bond_(literary_character).

the extraction by dictionary, we plan to make an automated regular expression generator that, given an entity, will match as many cases as possible. SPARQL queries using those seeds will then generate a dictionary composed of regular expressions that would match multiple derivation of the entities.

Acknowledgments. This work was primarily supported by the innovation activity PasTime (17164) of EIT Digital (https://www.eitdigital.eu).

References

1. Huang, P.-S., He, X., Gao, J., Deng, L., Acero, A., Heck, L.: Learning deep structured semantic models for web search using clickthrough data. In: 22nd ACM International Conference on Information & Knowledge Management (CIKM) (2013)
2. Manning, C.D., Surdeanu, M., Bauer, J., Finkel, J., Bethard, S.J., McClosky, D.: The stanford coreNLP natural language processing toolkit. In: Association for Computational Linguistics (ACL) System Demonstrations (2014)
3. Plu, J., Rizzo, G., Troncy, R.: A hybrid approach for entity recognition and linking. In: 12th European Semantic Web Conference (ESWC), Open Knowledge Extraction Challenge (2015)
4. Plu, J., Rizzo, G., Troncy, R.: Enhancing entity linking by combining NER models. In: 13th European Semantic Web Conference (ESWC), Open Knowledge Extraction Challenge (2016)
5. Plu, J., Rizzo, G., Troncy, R.: ADEL: adaptable entity linking. Semant. Web J. (SWJ) Spec. Issue Linked Data Inf. Extr. (2017)
6. Rizzo, G., van Erp, M., Troncy, R.: Inductive entity typing alignment. In: 2nd International Workshop on Linked Data for Information Extraction (LD4IE) (2014)
7. Speck, R., Ngomo, A
8. Usbeck, R., Ngomo, A., Röder, M., Gerber, D., Coelho, S., Auer, S., Both, A.: AGDISTIS - graph-based disambiguation of named entities using linked data. In: The Semantic Web–ISWC 2014 (2014)

Question Answering over Linked Data Challenge

7th Open Challenge on Question Answering over Linked Data (QALD-7)

Ricardo Usbeck[1(✉)], Axel-Cyrille Ngonga Ngomo[1], Bastian Haarmann[2], Anastasia Krithara[3], Michael Röder[1], and Giulio Napolitano[2]

[1] Data Science Group, Paderborn University, Paderborn, Germany
{usbeck,ngonga,roeder}@informatik.uni-leipzig.de
[2] Fraunhofer-Institute IAIS, Sankt Augustin, Germany
{bastian.haarmann,giulio.napolitano}@iais.fraunhofer.de
[3] National Center for Scientific Research "Demokritos", Athens, Greece
akrithara@iit.demokritos.gr

1   Introduction

The past years have seen a growing amount of research on question answering (QA) over Semantic Web data, shaping an interaction paradigm that allows end users to profit from the expressive power of Semantic Web standards while, at the same time, hiding their complexity behind an intuitive and easy-to-use interface. On the other hand, the growing amount of data has led to a heterogeneous data landscape where QA systems struggle to keep up with the volume, variety and veracity of the underlying knowledge.

The Question Answering over Linked Data (QALD) challenge aims at providing an up-to-date benchmark for assessing and comparing state-of-the-art-systems that mediate between a user, expressing his or her information need in natural language, and RDF data. It thus targets all researchers and practitioners working on querying Linked Data, natural language processing for question answering, multilingual information retrieval and related topics. The main goal is to gain insights into the strengths and shortcomings of different approaches and into possible solutions for coping with the large, heterogeneous and distributed nature of Semantic Web data.

QALD[1] has a 6-year history of developing a benchmark that is increasingly being used as standard evaluation tool for question answering over Linked Data. Overviews of the past instantiations of the challenge are available from the CLEF Working Notes as well as ESWC proceedings:

- QALD-6: http://www.springer.com/us/book/9783319465647
- QALD-5: http://ceur-ws.org/Vol-1391/173-CR.pdf
- QALD-4: http://ceur-ws.org/Vol-1180/CLEF2014wn-QA-UngerEt2014.pdf
- QALD-3: https://pub.uni-bielefeld.de/download/2685575/2698020

---

[1]http://www.sc.cit-ec.uni-bielefeld.de/qald/.

© Springer International Publishing AG 2017
M. Dragoni et al. (Eds.): SemWebEval 2017, CCIS 769, pp. 59–69, 2017.
https://doi.org/10.1007/978-3-319-69146-6_6

Furthermore, through the QALD challenge, we (1) provide objective measures for how well current systems perform on real tasks of industrial relevance and (2) detect bottlenecks of existing systems in order to further develop them and make them more usable in practice. Since many of the topics relevant for QA over Linked Data lie at the core of ESWC (Multilinguality, Semantic Web, Human-Machine-Interfaces), we have run the 7th instantiation of QALD again at ESWC 2017. This year the challenge was supported by the EU project HOBBIT [1], which has already established a network of people from the Semantic Web as well as the Big Data community, both from the academia and industries. In addition, HOBBIT provided an open source holistic benchmarking platform for Big Linked Data, in which the challenge was run. Thanks to the HOBBIT project we were able to guarantee a controlled setting involving rigorous evaluations via its platform.[2]

Similar Events. To the best of our knowledge, there is no event with a comparable scope (Linked, Large-Scale, Hybrid Data) outside this series in the Semantic Web Community. However, there has thus been a number of challenges and campaigns attracting researchers as well as industry practitioners to QA. Since 1998, the TREC conference, especially the QA track [4], aims at providing domain-independent evaluations over large, unstructured corpora as well as Community-based QA. Next to that, the BioASQ series [2] challenges semantic indexing as well as QA systems on biomedical data and is currently at its fifth installment. Here, systems have to work on RDF as well as textual data to present matching triples as well as text snippets. The OKBQA challenge[3] is primarily an open QA platform powered by several Korean research institutes but they also released the NLQ datasets.

2   Tasks and Datasets

The key challenge for QA over Linked Data is to translate a user's information need into such a form that it can be evaluated using standard Semantic Web query processing and inferencing techniques. The main task of QALD therefore is the following:

Given one or several RDF dataset(s) as well as additional knowledge sources and natural language questions or keywords, return the correct answers or a SPARQL query that retrieves these answers.

Data format

All data for the tasks can be found in our project repository https://github.com/ag-sc/QALD/tree/master/7/data. We encouraged the use of QALD-JSON

---

[2] https://project-hobbit.eu/challenges/qald2017/.
[3] http://www.okbqa.org.

format[4] as communication format between the systems and the GERBIL QA respectively HOBBIT platform:

```
1  {"id":"3",
2   "answertype":"resource",
3   "aggregation":false,
4   "onlydbo":true,
5   "hybrid":false,
6   "question":[
7      {
8        "language":"en",
9        "string":"Who was the wife of U.S. president Lincoln?",
10       "keywords":"U.S. president, Lincoln, wife"
11     },
12     {
13       "language":"nl",
14       "string":"Wie was de vrouw van de Amerikaanse president Lincoln?",
15       "keywords":"vrouw, president van America, Lincoln"
16     }
17   ],
18   "query":{
19     "sparql":"PREFIX dbo:<http://dbpedia.org/ontology/>
20         PREFIX res:<http://dbpedia.org/resource/>
21         SELECT DISTINCT ?uri
22         WHERE {res:Abraham_Lincoln dbo:spouse ?uri.}"
23   },
24   "answers":[
25     {
26       "head":{
27         "vars":[
28           "uri"
29         ]
30       },
31       "results":{
32         "bindings":[
33           {
34             "uri":{
35               "type":"uri",
36               "value":"http://dbpedia.org/resource/Mary_Todd_Lincoln"
37             }
38           }
39         ]
40   }}]}
```

In order to focus on specific aspects and challenges, we included the following four tasks.

Task 1: Multilingual Question Answering over DBpedia. Given the diversity of languages used on the web, there is an increasing need to facilitate multilingual access to semantic data. The core task of QALD is thus to retrieve answers from an RDF data repository given an information need expressed in a variety of natural languages.

---

[4]https://github.com/AKSW/gerbil/wiki/Question-Answering and the results are formatted according to https://www.w3.org/TR/sparql11-results-json/.

Training data. The underlying RDF dataset was DBpedia 2016-04. The training data consists of 215 questions compiled and curated from previous challenges. The questions are be available in eight different languages (English, Spanish, German, Italian, French, Dutch, Romanian and Farsi). Those questions are general, open-domain factual questions, for example:

(en) *Which book has the most pages?*
(de) *Welches Buch hat die meisten Seiten?*
(es) *¿Que libro tiene el mayor numero de paginas?*
(it) *Quale libro ha il maggior numero di pagine?*
(fr) *Quel livre a le plus de pages?*
(nl) *Welk boek heeft de meeste pagina's?*
(ro) *Ce carte are cele mai multe pagini?*

The questions vary with respect to their complexity, including questions with counts (e.g., *How many children does Eddie Murphy have?...*), superlatives (e.g., *Which museum in New York has the most visitors?*), comparatives (e.g., *Is Lake Baikal bigger than the Great Bear Lake?*), and temporal aggregators (e.g., *How many companies were founded in the same year as Google?*). Each question is annotated with a manually specified SPARQL query and answers. In the above case, the SPARQL query looks as follows:

```
SELECT DISTINCT ?uri
WHERE {
    ?uri a <http://dbpedia.org/ontology/Book> .
    ?uri <http://dbpedia.org/ontology/numberOfPages> ?n .
}
ORDER BY DESC(?n)
OFFSET 0 LIMIT 1
```

And the answer is <http://dbpedia.org/resource/The_Tolkien_Reader>.

Test Data. The test dataset consists of 50 similar manually created questions. However, this year we decided to increase the complexity of the test data and add several other question types including questions according to RDF types (e.g., *What is backgammon?...*) or questions demanding mathematical operations (e.g., *What is the radius of the earth?...*). They are compiled from existing, real-world question and query logs, in order to provide unbiased questions expressing real-world information needs. The questions were manually curated to ensure a high quality standard.

Task 2: Hybrid question answering

A large amount of information is still available as unstructured text only, both on the web and in the form of labels and abstracts in Linked Data sources. Therefore, approaches are needed that can not only deal with the specific character of structured data but also with finding information in other sources, processing both structured and unstructured information, and combining such gathered information into a single answer. Therefore, QALD-7 included a task on hybrid question answering, forcing systems to retrieve answers for questions that required the integration of data both from RDF and from textual sources.

Training data. The training data is build using DBpedia 2016-04 as the RDF knowledge base, together with the English Wikipedia as the textual data source. As training data, we included 105 questions in English from past challenges (partly based on questions used in the INEX Linked Data track[5]). The questions are annotated with answers as well as a pseudoquery that indicates what information can be obtained from RDF data and what from free text. The pseudoquery is like an RDF query but may contain free text as subject, property or object of a triple. An example is the question *Who is the front man of the band that wrote Coffee & TV?*, with the following corresponding pseudoquery:

```
SELECT DISTINCT ?uri
WHERE {
    <http://dbpedia.org/resource/Coffee_&_TV>
    <http://dbpedia.org/ontology/musicalArtist> ?x .
    ?x <http://dbpedia.org/ontology/bandMember> ?uri .
    ?uri text:"is"text:"frontman" .
}
```

The manually specified answer is <http://dbpedia.org/resource/Damon_Albarn>.

Test data. As test questions, we generated 50 similar questions, all manually created and checked by at least two data experts. The main goal in devising those questions was not to take into account the vast amount of data available and the problems arising from noisy, duplicate and conflicting information. Rather, we aimed at enabling a controlled and fair evaluation, considering that hybrid question answering is still a very young line of research.

Task 3: Large-Scale Question answering over RDF

A new task was introduced this year, with focus on large-scale question sets. The aim was to assess approaches able to scale up to a big data volume, handle a vast amount of questions and speed up the question answering process by parallelization, such that the highest possible number of questions can be answered as accurately as possible in the shortest possible time. Again, the data for this task is based on the DBpedia 2016-04 RDF knowledge base.

Data creation. The training set consists of 100 questions compiled from the HOBBIT project. The test set of 2M questions is generated by an algorithm deriving new questions from the training set by varying both the query desire and the form of the natural language expression. Questions were annotated with SPARQL queries and answers and, in the test scenario, they were sent every

---

[5]http://inex.mmci.uni-saarland.de/tracks/dc/index.html.

minute to the competing systems in packets of increasing size, with $n+1$ questions asked at minute n. Participating systems were evaluated with respect to both number of correct answers and time needed.

Task 4: Question Answering over Wikidata. This task, also new for the 7th edition of the QALD challenges, provided a benchmark focusing on the ability of systems to adapt to new data sources. Questions originally formulated for DBpedia require an answer using Wikidata, so that systems have to deal with a different data representation structure. The task is meant to support the evaluation of how generic the approach of a given system is and how easy it is to adapt to a new data source.

Data creation. The training question set consisted of 100 questions selected from Task 1 of the QALD-6 challenge. We formulated the queries to answer these questions from Wikidata and generated the gold standard answers using them on the Wikidata dump from 09-01-2017. As test data, 50 additional questions were used from the QALD-6 challenge (Table 1).

Table 1. Number of questions per task in the training and test sets.

Task	Train	test
1. Multilingual	215	50
2. Hybrid	105	50
3. Large-scale	100	2M
4. Wikidata	100	50

3 Evaluation

The QALD challenge provides an automatic evaluation tool (GERBIL QA [3] integrated into the HOBBIT platform)[6, 7] that is open source and available for everyone to re-use. The GERBIL QA platform is accessible online, so that participants can simply upload the answers produced by their system or even check their system via a webservice. Each experiment has a citable, time-stable and archivable URI which is both human- and machine-readable. However, participating systems had to provide a Docker container[8, 9] to participate in the final challenge which communicated with the HOBBIT platform.

---

[6] http://gerbil-qa.aksw.org/gerbil/.
[7] http://project-hobbit.eu/.
[8] https://project-hobbit.eu/challenges/qald2017/#Technical_requirements.
[9] https://github.com/hobbit-project/platform/wiki/Participate-in-a-challenge.

The QA systems were evaluated with respect to precision and recall. For each question q, precision and recall are computed as follows:

$$\mathrm{recall}(q) = \frac{\text{number of correct system answers for } q}{\text{number of gold standard answers for } q}$$

$$\mathrm{precision}(q) = \frac{\text{number of correct system answers for } q}{\text{number of system answers for } q}$$

The evaluation computed the macro and micro F-measure of a system over all test questions. That is, for micro F-measure we summed up all true and false positives and negative up and calculated in the end the precision, recall and F-measure while for the macro measures we calculated precision, recall and F-measure per question and averaged the values in the end.

In this challenge, we left out the computation of measures over only those questions that the system did provide an answer for. That is, question without answer would have been ignored instead of resulting in a zero F-measure. Thus, it was not possible to make an evaluation which would have allowed to take into account a system's ability to identify questions that it cannot answer. For task 3, specifically, the evaluation takes into account not only the accuracy measures for the answered questions but also the scalability measures in terms of number of processed queries and time needed for answer retrieval.

4   Participating Systems

Three teams participated in the QALD-7 challenge, with three teams addressing the multilingual task (two for English and one French) and two addressing the QA over Wikidata task. Note, that the description of the papers can be found in the challenge proceedings of the 2017 ESWC satellite proceedings.

WDAqua is a rule-based system using a combinatorial approach to generate SPARQL queries from natural language questions, leveraging the semantics encoded in the underlying knowledge base. It can answer questions on both DBpedia (supporting English) and Wikidata (supporting English, French, German and Italian). The system, which does not require training, participated in Task 1 and 4 of the challenge.

AMAL has been developed for QA in French. Firstly, the question type (e.g. *Boolean* or *Entity*) is classified by pattern matching. This induces the rerouting to the relevant question type solver where entities and properties are extracted: the former by syntactic parsing and subsequent linking to DBpedia entities; the latter by removing the found entity and searching for corresponding properties in DBpedia, possibly with the help of with Wikipage disambiguation links. SPARQL predicate identification is supported by a manually curated lexicon of common DBpedia properties, each linked to one or more possible French expressions. The system can only answer simple questions (concerning a single entity or a single property of an entity) and participated in Task 1.

Sorokin and Gurevych participated in Task 4 of the challenge. They provided a system producing the semantic representation of a natural language question, which is then deterministically converted into SPARQL. After minimal pre-processing, including POS tagging and entity linking, an end-to-end neural architecture employs a CNN neural scorer to choose among multiple semantic representations of the question. First, the semantic representations are generated by expansion on the knowledge base, guided by the entity found in the question and by all possible relations and constraints as present in the KB for the entity. Then, each question and candidate representations are vectorialised, with the CNN producing comparison scores based on cosine similarity, leading to the final choice.

ganswer2 [5] has participated outside the actual challenge this year as a system without a paper submission in Task 1. Zou et al. use a graph-based approach to generate a semantic query graph which reduced the transformation of natural language to SPARQL to a subgraph matching problem.

5   Results

Task 1: Multilingual question answering over DBpedia

Task 1 was run for the seventh time in 2017. Three participating teams submitted their systems via the HOBBIT or GERBIL QA platform. Please note that AMAL submitted their results as files, due to constraints of the system which resulted in using GERBIL QA as a platform. Also note that WDAQUA macro values are taken from the system authors challenge submission, as we could not use the platform at the time of this publication. Furthermore, we used GERBIL QA for the training data evaluation as HOBBIT was only targeted for the evaluation of the actual challenge (blind test) data.

The experimental data for task 1 over training data can be found in the following:

- ganswer   (en):   http://gerbil-qa.aksw.org/gerbil/experiment?id=20170630 0001,
- AMAL (fr): http://gerbil-qa.aksw.org/gerbil/experiment?id=201706300002.

The experimental data for task 1 over the test data can be found in the following:

- ganswer   (en):   http://master.project-hobbit.eu/#/experiments/details? id=1498647986590,
- WDAqua   (en):   http://master.project-hobbit.eu/#/experiments/details? id=1498647742687,
- AMAL (fr): http://gerbil-qa.aksw.org/gerbil/experiment?id=201706300011.

By providing human- and machine-readable experimental URIs, we provide deeper insights and repeatable experiment setups.

Note also that the numbers reported here may differ from the publications of the participants, as these figures were not available at the time of participant paper submission (Table 2).

Table 2. Overview over QALD-7 task 1.

Test	WDAqua	ganswer2	AMAL
Language	en	en	fr
Error count		3	
Micro Precision	0.080	0.322	0.998
Micro Recall	0.006	0.127	0.989
Micro F1-measure	0.012	0.182	0.993
Macro Precision	0.162	0.487	0.720
Macro Recall	0.160	0.498	0.720
Macro F1-measure	0.143	0.469	0.720
Train	WDAqua	ganswer2	AMAL
Language	en	en	fr
Error count			
Micro Precision	-	0.113	0.971
Micro Recall	-	0.561	0.697
Micro F1-measure	-	0.189	0.811
Macro Precision	0.490	0.557	0.750
Macro Recall	0.540	0.592	0.751
Macro F1-measure	0.510	0.556	0.751

Task 4: Question answering over Wikidata

Task 4 was run this year at QALD-7 for the first time and announced at short notice. Thus, it only attracted two teams. However, both teams performed well on both the train and the test datasets. For the first time in QALD, Sorokin and Gurevych also used a neural network to answer questions over Wikidata. As can be seen from the numbers in Table 3, both systems have a higher macro F-measure than micro F-measure. The task 4 data contains questions with long answer lists and if a system fails to answer such querys this has a huge impact on its micro recall and thus on its micro F-measure.

Table 3. Overview over QALD-7 task 4. The experimental data for task 4 executed with the HOBBIT platform can be found here http://master.project-hobbit.eu/#/experiments/details?id=1498647794373, http://master.project-hobbit.eu/#/experiments/details?id=1498647917506, http://master.project-hobbit.eu/#/experiments/details?id=1498647883035 and http://master.project-hobbit.eu/#/experiments/details?id=1498647941734.

Test dataset	WDAqua	Sorokin and Gurevych
Micro Precision	0.392	0.428
Micro Recall	0.082	0.030
Micro F1-measure	**0.136**	0.057
Macro Precision	0.739	0.661
Macro Recall	0.606	0.430
Macro F1-measure	**0.552**	0.427
Train dataset	WDAqua	Sorokin and Gurevych
Micro Precision	0.172	0.295
Micro Recall	0.112	0.070
Micro F1-measure	**0.136**	0.113
Macro Precision	0.759	0.774
Macro Recall	0.710	0.756
Macro F1-measure	0.636	**0.645**

6   Summary

The seventh Question Answering over Linked Data challenge introduced two new tasks (scalable QA and QA over Wikidata) and repeated two of the successful past tasks. For the first time, the participating systems offered webservices as a prerequisite to participate in the challenge which will support comparable research in the future. In this challenge, we also changed the underlying evaluation platform to account for the need for comparable experiments via webservices and new technologies such as docker as compared to former XML/JSON file submissions. This increased the entranced barrier for participating teams but ensures a long term comparability of the system performance and a fair and open challenge.

In the future, we will further simplify the participation process and offer leader boards prior to the actual challenge in order to allow participants to already see their performance. After feedback from the authors, we will add new key performance indicators to also account for the capability of a system to know which questions it cannot answer and take confidence scores for answers into account. Overall, we hope that the HOBBIT platform can serve as a long term challenge support to increase comparable and repeatable question answering research.

Acknowledgments. This work was supported by the Eurostars projects DIESEL (E!9367) and QAMEL (E!9725) as well as the European Union's H2020 research and innovation action HOBBIT under the Grant Agreement number 688227. We also want to thank Christina Unger and Sebastian Walter for supporting this challenge.

References

1. Ngomo, A.-C.N., García-Rojas, A., Fundulaki, I.: HOBBIT: holistic benchmarking of big linked data. ERCIM News **2016**(105) (2016)
2. Tsatsaronis, G., Balikas, G., Malakasiotis, P., Partalas, I., Zschunke, M., Alvers, M.R., Weissenborn, D., Krithara, A., Petridis, S., Polychronopoulos, D., Almirantis, Y., Pavlopoulos, J., Baskiotis, N., Gallinari, P., Artières, T., Ngonga, A., Heino, N., Gaussier, É., Barrio-Alvers, L., Schroeder, M., Androutsopoulos, I., Paliouras, G.: An overview of the BIOASQ large-scale biomedical semantic indexing and question answering competition. BMC Bioinformatics **16**, 138 (2015)
3. Usbeck, R., Michael, R., Unger, C., Hoffmann, M., Demmler, C., Huthmann, J., Ngomo, A.-C.N.: Benchmarking question answering systems. Technical report, Leipzig University (2016)
4. Voorhees, E.M., et al.: The trec-8 question answering track report. Trec **99**, 77–82 (1999)
5. Zou, L., Huang, R., Wang, H., Yu, J.X., He, W., Zhao, D.: Natural language question answering over RDF: a graph data driven approach. In: Proceedings of the 2014 ACM SIGMOD international conference on Management of Data, pp. 313–324. ACM (2014)

End-to-End Representation Learning
for Question Answering with Weak Supervision

Daniil Sorokin[(✉)] and Iryna Gurevych

Ubiquitous Knowledge Processing Lab (UKP-TUDA)
Department of Computer Science,
Technische Universität Darmstadt, Darmstadt, Germany
sorokin@ukp.informatik.tu-darmstadt.de
http://www.ukp.tu-darmstadt.de

Abstract. In this paper we present a knowledge base question answering system for participation in Task 4 of the QALD-7 shared task. Our system is an end-to-end neural architecture for constructing a structural semantic representation of a natural language question. We define semantic representations as graphs that are generated step-wise and can be translated into knowledge base queries to retrieve answers. We use a convolutional neural network (CNN) model to learn vector encodings for the questions and the semantic graphs and use it to select the best matching graph for the input question. We show on two different datasets that our system is able to successfully generalize to new data.

Keywords: Semantic web · Question-answering · Representation learning · Convolutional neural networks · Semantic parsing · Weak supervision

1 Introduction

QALD is a series of international competitions on mapping natural language questions to knowledge base queries [17]. The goal of the competitions is to provide a benchmark for natural language based interfaces to knowledge bases.

In this paper, we present a system that was developed for Task 4 of the QALD-7 shared task, "English question answering over Wikidata". The task is formulated as follows: given a natural language question, translate it into a structured query in SPARQL that can be executed against Wikidata to obtain the answer to the question. The provided training data set for Task 4 consists of 100 natural language questions, the answers may be real word entities, numbers or dates. Wikidata [19] is a popular collaboratively constructed knowledge base that contains around 17 million entities and more than 70 million facts of common knowledge.

In our system, we implement a semantic parsing approach to the problem of knowledge base question answering (factoid QA). That is, we produce semantic representations for natural language questions that are then deterministically converted into SPARQL queries and executed against Wikidata.

© Springer International Publishing AG 2017
M. Dragoni et al. (Eds.): SemWebEval 2017, CCIS 769, pp. 70–83, 2017.
https://doi.org/10.1007/978-3-319-69146-6_7

Multiple successful question answering systems were presented in the previous QALD competitions [17], as well as in conjunction with other QA datasets [3,13,16]. The key challenge in this respect is how to encode the semantics of the question and to use it to find the correct answer. This can be done either by directly encoding the question meaning into a latent vector encoding (end-to-end systems) or by constructing an explicit structural semantic representation (semantic parsing systems). The latent vector representation is normally used to score individual answer candidates contained in the KB [7,8,11], whereas the structural semantic representation is converted to a query to be executed against the KB [3,22].

Semantic parsing systems, such as [2,3,9], usually relied on trained models with manually defined features and therefore, suffer from error propagation [15]. End-to-end systems that learn latent vector encodings for questions and answers eliminate this problem [8,13]. However, latent vector encodings are hard to analyze for errors or to modify with explicit constraints. Questions that require aggregation over several knowledge base entities or temporal constraints are almost impossible to model with the current end-to-end models (see, for example, error analysis in [8]).

In our approach, we combine the best of the latent vector encodings and explicit semantic representation methods. Our main contribution is an end-to-end iterative generation of multi-relational semantic representations that integrates a neural network to learn vector encodings for questions and semantic representations. We use the similarity between the vector encodings to choose the correct semantic representation for a given question.

The end-to-end neural architecture doesn't need handcrafted features or heavy pre-processing that are required in other approaches. It automatically learns a correspondence between structural and lexical features of a semantic representation and a natural language question. Thus, our approach can better generalize to new unseen questions than approaches based on manually defined features and can directly integrate explicit constraints.

We demonstrate the effectiveness of our system on two datasets: QALD-7 Task 4 and WebQuestions [3]. Both dataset contain questions that require complex reasoning to be answered.

2   Related Work

The existing semantic parsing approaches to knowledge base question answering usually consist of a mechanism that generates acceptable semantic representations and of a model that relies on a combination of hand-crafted features to select the correct representation [3,16,22]. As opposed to the end-to-end approaches, error propagation is the main downside of the semantic parsing solutions. For example, Reddy et al. [15] estimate that over 35% of errors are being propagated down the pipeline. We try to overcome this in our approach by designing an end-to-end architecture to process the semantic graphs. Dong et al. [8] and Jain [11] achieve the best results on factoid QA with an end-to-end

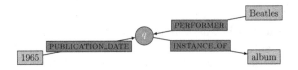

Fig. 1. Graphical semantic representation for a question "What albums did the Beatles release in 1965?"

approach and innovative usage of neural networks to search through the KB. However, their approaches don't use explicit semantic representations and thus fail on cases when explicit constraints are required.

Encoding the semantics of a questions using semantic graphs is a common way to conceptualize semantic representations [3,15,22]. Our graphs are most similar to those of [15,22]. We closely follow the approach of [22] who, in contrast to [15], don't rely on syntactic parsing to construct semantic graphs. At the same time, our approach is more flexible than [22] because we don't separate out a single main relation and we are able to process all relations in the same way. This is mainly possible because we are using Wikidata that uniformly encodes all information with binary relations.

A different approach was taken in one of the winning systems of QALD-6 [9]. The authors have used a controlled language to enforce restrictions on syntax and lexical content of a question. This has allowed to unambiguously map the question to a semantic representation and retrieve answers with high precision. The system has demonstarted a very high performance on questions form a closed domain, but wouldn't be able to answer if a question is not covered by the controlled language.

A set of QA system exists that exclusively focus on questions that can be answered using a single triple from the KB [5,13]. These systems don't incorporate constraints or multi-relational representations and usually model the task as a classification problem. Given a question, one has to predict a relation type from a pre-defined scheme. We don't compare to these approaches, since our focus is on complex questions.

3 Semantic Graphs

We use a graphical representation to encode the semantics of a questions (*semantic graph*). Our semantic representations (see Fig. 1) consist of a question variable node (q), real world entities (Beatles), constraints (argmin) and relation types from the KB (PERFORMER). The question variable denotes the answer to the question. That is, all entities from the KB that can take its place so that all relations and constraints hold, constitute the answer to the question.

To retrieve the answers given a semantic graph, we convert it to a SPARQL query. All relations in the semantic graph are directed and the conversion is straightforward. We add an ORDER BY clause if there is a temporal constraint in the graph. The query is executed against a Wikidata RDF dump that is stored locally in Virtuoso[1]. To speed up the query, we blacklist certain relations and entities that are used to encode meta-information in Wikidata.

All relation are attached to the question variable node and we don't allow anything but a Wikidata entity in the position of the question variable. This poses limitations on the types of questions that our system can answers (e.g., aggregate questions or true/false question won't be processed), but it also limits the space of possible graphs and makes the search for the best matching graph more tractable.

Our semantic graphs are coupled to the knowledge base and therefore, only relations and entities defined by the knowledge base scheme are possible. In the following sections, we describe the way we construct semantics graphs for a given question and how we select the graph that matches the semantics of the question the best.

4   System Architecture

4.1   Entity Linking

Our system takes a natural language question in the form of a string as input. We tokenize it and add part-of-speech tags with the Stanford CoreNLP toolkit [14]. Afterwards, we extract token fragments using a set of regular expression rules that match all sequences of nouns with adjacent modifiers. For each extracted fragment, we generate a set of possible token n-grams and look them up in Wikidata. That gives a list of Wikidata entities that might correspond to the given fragment.

Since Wikidata doesn't offer an entity linking API, we have used alternative labels of the Wikidata entities to perform the look-up. Alternative labels are entered manually for each Wikidata entity and represent different spelling and name variations. For example, the entity *album*:Q482994 has the following alternative labels: [*audio album, music album, record album*].

Following the approach in [1], we sort the retrieved list of entities by the combination of the Levenshtein distance between the fragment and the item label and the integer part of the item ID:

$$
\begin{aligned}
rank =& a \text{ levenshtein}(\textit{fragment}, \textit{entity\_main\_label}) \\
&+ b \log \textit{entity\_serial\_id} \\
&+ c \max\left(1 - \frac{\text{len}(\textit{entity\_label})}{\text{len}(\textit{fragment})}, 0\right)
\end{aligned}
\tag{1}
$$

Since in some cases only a part of a fragment will match an entity, we also add a term to prefer longer fragment matches. The coefficients a, b and c were heuristically set to 1, 1 and 2 respectively. We select the candidate with the smallest rank for each fragment as the final linking.

---

[1] https://virtuoso.openlinksw.com.

For example, in the question "What was the first album released by the Beatles?", we first extract fragments "the first album" and "the Beatles" and then link them to entities *The Beatles (band)*:Q1299 and *album (musical record)*:Q482994.

4.2 Iterative Representation Generation

Once the list of entities is extracted from the question, we use it to construct possible semantic graphs. We develop a representation generation procedure that defines what kind of graphs can be constructed.

We iteratively generate candidate semantic graphs of the question using a set of actions which can be applied at each step, starting with an empty graph that contains only a question variable. We define three types of actions for graph generation: ADD_RELATION, ADD_TEMPORAL_CONSTRAINT, ADD_NUMBER_CONSTRAINT. The actions define how we search for possible semantic representations. Each action creates a new modified copy of the graph and adds to the list of candidates. Our procedure is inspired by the process of adding constraints to the question in [2], yet our approach is more flexible because we don't divide the representation into the main relation and constraints. Figure 2 shows the application order of the actions.

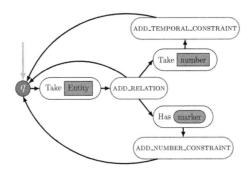

Fig. 2. Scheme of steps that can be undertaken to construct a graph.

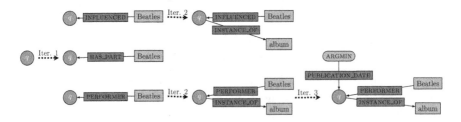

Fig. 3. Generating candidate representations for "What was the first Beatles album?"

For each action, we define conditions that must be satisfied in order for the action to be applied at the current step. We list the conditions for each action in Table 1. The conditions control the flow of the graph generation procedure. For example, at the first iteration in Fig. 3 we apply the ADD_RELATION action, since it is the only action that can be performed on a empty graph. The result is one graph for each relation that exists for the entity $\boxed{\text{Beatles}}$ (Fig. 3 shows only three). It is followed by another application of ADD_RELATION since there is a second entity in the question and finally, ADD_TEMP_CONSTRAINT can be applied at the third iteration step because of a temporal marker "first" in the question. We check that each candidate semantic graph is valid and those that don't produce answers are not further expanded. For example, in Fig. 3 we don't expand the candidate in the middle after the first iteration, since it is impossible to add a relation with $\boxed{\text{album}}$ that would result in a valid semantic graph.

Table 1. The list of actions defined for the iterative representation generation process (E–list of entities, Q–list of question tokens, s–current semantic representation)

Action	Conditions	Action description
ADD_RELATION	LEN$(E) > 0$	Queries Wikidata for relations R that exist for $e, e \in E$, and creates a new representation for each $r, r \in R$
ADD_TEMPORAL _CONSTRAINT	LEN(RELATIONS$(s)) > 0 \wedge$ LEN($temp\_markers \cap Q) > 0$	Creates a new representation with a constraint that the answer is the last or the first entity in a temporally sorted list
ADD_NUMBER _CONSTRAINT	LEN(RELATIONS$(s)) > 0 \wedge$ CONTAINS$(Q, number)$	Creates a new representation with an added relation that has a numeric argument (e.g. year)

4.3 Neural Vector Encodings

We construct a neural network model to select the best matching semantic graph for the question. It encodes the question and the candidate semantic graphs into fixed-size vectors and then uses the cosine measure to find the correct graph. The semantic graph that has the closest vector to the question vector is taken to be the best semantic representation of the question.

The end-to-end architecture jointly learns vector encodings for questions and semantic graphs. We use the same CNN-based model to encode both the question and the individual relations of the semantic graph. The encodings of the individual relations are later composed into a single vector for the whole graph. We choose CNNs as a basis for our neural network model, since they have proven to be successful for question answering [2,8].

The architecture of the model is represented in Fig. 4, where it is used to encode an example question into a fixed-size vector. The input question is first tokenized and the tokens corresponding to named entities are replaced with a special $\langle e \rangle$ token. We also mark the beginning and the ending of the input with $\langle S \rangle$ and $\langle E \rangle$ respectively. The resulting list of tokens $\mathbf{x} = \{x_1, x_2 \ldots x_n\}$ constitutes the input to the model (see at the bottom of Fig. 4).

Next, we represent each token as a list of its character trigrams using the hashing technique suggested in [10]. For example, the word "what" has the following trigrams: $\mathbf{t} = \{\#wh, wha, hat, at\#\}$, where $\#$ stands for the word boundary. The word is represented as a binary vector $\mathbf{h} \in \mathbb{R}^{|V|}$, where V is the number of possible trigrams in the training data. For the word "what", we mark the positions that correspond to the trigrams in $\mathbf{t}$ with 1 and the rest is 0. Such scheme ensures that different morphological forms of the same word or misspelled words have a similar representation. In the preliminary experiments, we have also observed that this scheme performs more consistent and better than using word or character embeddings.

The list of token representations is further processed by the CNN layer C. For each token, it convolves its representation with the representation of the neighboring tokens. We apply the max pooling operation after the CNN layer to capture the most salient features of the input string. The output of the max pooling operation c is further transformed with a fully connected layer $\mathbf{S}$ and a tanh non-linearity. We take the resulting vector $\mathbf{s}_q$ as the latent encoding of the question.

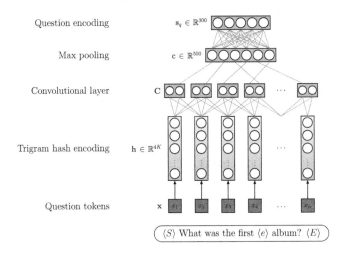

Fig. 4. The architecture of the CNN-based question encoder

To encode a semantic graph, we first break it into individual relations. For each relation, we construct a string label by taking the Wikidata relation type label and adding the $\langle e \rangle$ token either at the beginning or the end depending on the direction of the relation. For temporal constraints, we always use the label "point in time" and a $\langle a \rangle$ token instead.

We tokenize the relation labels and use them as an input to the same CNN-based model that was used to encode the question (see Fig. 5). The output is a semantic vector for each individual relation in the graph: $\{\mathbf{s}_{r_1}, \mathbf{s}_{r_2} \ldots \mathbf{s}_{r_m}\}$. The weights of the neural network model are shared in both cases and the vector encodings for questions and semantic graphs are learned jointly. To get a single vector for the whole graph $\mathbf{s}_g$, we apply another max pooling operation on the set of the relation vectors. The order of relations in the graph is not important and the max pooling disregards the order of input elements. The final vector encoding for a candidate graph encodes the most prominent features of the relations that it contains.

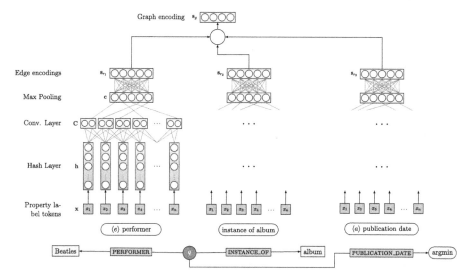

Fig. 5. Graph encoder architecture, here used to encode the example graph in Fig. 1.

5 Question Answering as Graph Generation

In this section, we describe how the system proceeds to answer a given question. First, the input question is encoded into a vector with the question encoder (Fig. 4) and the question vector encoding is stored for further reference. Second, we extract entities from the question to start the graph construction. We take the steps described in Sect. 4.2 to construct possible semantic graphs for the input question. Each constructed variant is encoded by processing individual semantic relations in the graph and combining them into a single graph encoding (Fig. 5).

Finally, we score each semantic graph using the cosine distance between the vector encoding of the question and the vector encoding of the graph. During evaluation we perform a beam search and score the constructed graphs at each step, selecting the top 10 graphs for further processing. The semantic graph with the highest score is selected as the final choice for the given question and is used to retrieve the answers from the KB.

6   Model Training

To train the model we need positive pairs of questions and semantic graphs. We use weak supervision in the form of question-answer pairs as suggested in [3] to train the neural network model. Weak supervision can provide more training data than available in the form of manually annotated semantic representations. We take the training subset of the WebQuestions dataset [3] which contains 3778 questions and manually retrieved answers. To get pairs of questions and semantic graphs for model training, we run our graph generation procedure on each question. We evaluate each possible semantic graph against Wikidata and compare the extracted answers to the manually provided answers in the dataset. The semantic graphs that result in F1 higher than a certain threshold are stored as positive training instances and the rest of the graphs generated during the same process are used as negative instances. We set the threshold to 0.2 to capture as much of the positive semantic graphs as possible. To increase the search space, we additionally allow second-order relations at this step.

Since WebQuestion was originally developed for the Freebase knowledge base, not all of the questions in the dataset can be answered with Wikidata. With our method, we generate positive semantic graphs for 2334 question from the training part of WebQuestions and reserve 702 of them for validation.

At each training epoch we take all positive semantic graphs and sample up to 20 negative graphs per question. We use the respective F1-scores of the positive semantic graphs to define the training objective. We apply the softmax transformation on the list of F1-scores of the positive graphs and the sampled negative graphs and use the Kullback-Leibler divergence as the loss function. This proves to be effective, since many questions have multiple positive graphs, none of which achieve a perfect F1-score.

The loss is computed for each training instance (a question and a set of semantic graphs) and is averaged over a batch of size 128. The Adam optimizer [12] is used to perform the updates on the weights of the network. We determine the rest of hyper-parameters with the random search on the validation set. We set the filter length of the CNN to 3 and the step size is set to 1. The size of the CNN layer output is 500 and the dimension of the question and graph vector encodings is 300.

7   Experiments

In Table 2, we report preliminary evaluation results on the training dataset for Task 4 of the QALD-7 Shared Task using the metrics from [17]. Our model was not trained on this dataset and, therefore, the reported results represent an expected generalization error of our system.

As mentioned in Sect. 3, the system currently doesn't cover the questions that require a number or a year as an answer. Therefore, only 80 out of 100 Task 4 dataset questions could be processed by our system.

Table 2. Evaluation results on the QALD-7 Task 4 training (100 questions)

	Processed	Right	Partially	Precision	Recall	F1	G. F1
WDAqua (full) [6]	100			0.320	0.323	0.322	0.322
WDAqua (keywords)[6]	100			0.280	0.280	0.280	0.280
Our system	80	25	36	0.351	0.432	0.364	0.291
Our system (ideal model)	80	47	30	0.760	0.898	0.727	0.581

For comparison, we list results for a competitor system, WDAqua [6], on the same dataset (see the paper for the description of the system). These are the only other results that were published on the QALD-7 Task 4 dataset so far. As opposed to our system, WDAqua can produce numbers and boolean values as answers, but it only allows for a maximum of two relations in a question and doesn't support superlative constructions. Our system proofs to be more flexible and outperforms WDAqua on precision, recall and F1 metrics.

Additionally, we include a version of our system with an oracle neural network model, that always chooses the correct semantic graph. This demonstrates the limitation of our semantic graphs, as the oracle system only achieves an F1 of 0.727. Right now, the semantic graphs don't cover questions that require complex semantic representations and comparison functions. Therefore, questions such as "Show me all basketball players that are higher than 2 meters." could be only partially answered.

Table 3. Evaluation results on the WEBQUESTION dataset

		Prec.	Rec.	F1
No pre-training	Yao and Van Durme (2015) [21]	0.372	0.596	0.422
	Berant et al. (2013) [3]	0.521	0.591	0.534
	Berant and Liang (2014) [4]	0.550	0.601	0.561
	Yao (2015) [20]	0.565	0.761	0.603
	Our system	0.604	0.638	0.610
	Reddy et al. (2016) [16]	0.663	0.750	0.679
Systems with pre-training	Yih et al. (2015) [22]	0.670	0.815	0.698
	Jain (2016) [11]	0.693	0.853	0.725

To directly compare our system to related work, we also perform an evaluation on the test subset of the WebQuestions dataset. It contains simple questions that can be answered with a single relation as well as complex questions that require multiple relations and constraints. WebQuestion has been a common benchmark for semantic parsers and information retrieval systems for many years.

A system's performance on WebQuestions is measured using *precision, recall* and *F1-score*. That ensures a fair evaluation, since a system might provide a partially correct answer that is nevertheless better than a complete miss.

Table 3 summarizes our results on the test part of WebQuestions. We evaluate on a subset of the test set that is substantially covered by Wikidata. We define this subset by searching through the space of possible semantic graphs of arbitrary depth to find questions that can be answered with Wikidata. Practically, this amounts to evaluating if a property path exists between entities in the question and the answers. We retain the questions that have a Wikidata answer with an F1-score higher than 0.8, which results in a subset of 460 questions for evaluation. We compute the results on this subset for other systems that were previously evaluated on the WebQuestions dataset using the systems' output posted by the authors.

As can be seen, our system compares favorably to the rest of the published results outperforming 4 out of 7 systems. 2 out of 3 systems that score better than our approach, [11,22], use unsupervised and semi-supervised pre-training on large web corpora such as ClueWeb. Yih et al. [22] additionally employ an entity linking system that is not openly available. They note that their system's performance drops by more than 8% when using alternatives. Reddy et al. [16] don't use unsupervised pre-training, but rely on a deprecated Freebase API for entity linking and make a heavy use of syntactic pre-processing that is not required for our approach. It is important to note that our system currently doesn't use any additional training data or unsupervised pre-training, but the same techniques can be used to improve our approach as well. We leave this directon for future work.

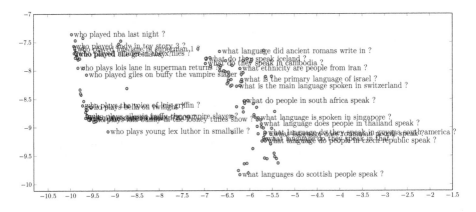

Fig. 6. Two clusters of question encodings learned by the model, every 5th question is labeled. Clusters corresponding to questions about an official language of a country and an actor's role are visible.

8   Model Analysis

Our architecture is able to learn fixed-size vectors for question and semantic graphs. In this section, we briefly analyze the vector encodings for questions that were learned by our model. We take the multi-dimensional encodings and map them into 2-D space using T-SNE [18] to be able to inspect them visually. We use our training dataset for the analysis, since it contains many question of similar semantics.

Figure 6 shows some of the clusters that can be identified among the question encodings. There we can see in detail that formed clusters correspond to questions with similar meaning. The left cluster consists of questions that ask about various character roles in movies. The right cluster groups questions that are concerned with languages spoken in a particular country.

It can been seen that the model learns to group questions with similar meanings but no obvious lexical overlap. For example, questions "What is the primary language of Israel?" and "What do people speak in South Africa speak?" both appear in the center of the right cluster on Fig. 6. Some errors are also evident from the diagram: for example, the question "Who played NBA last night?" is incorrectly placed near the character-role-cluster because of the second meaning of the word "play".

9   Conclusions

In this paper, we have presented an end-to-end system that produces semantic representations for natural language questions and evaluates them on Wikidata. We have demonstrated the soundness of our approach by comparison with other systems on two different QA datasets. Our system produces Wikidata items as answers and can successfully process more than 50% of the questions in the QALD-7 Task 4 dataset. On a popural WebQuestion dataset, our system shows the strongest results among the systems that don't rely on semi-supervised or unsupervised pre-training.

There are several obvious directions for future work that we hope to pursue. First, the unsupervised pre-training seems to be a logical way to improve the performance of our system. Second, as observed in [22], entity linking can have a big impact on the overall performance. Our approach to entity liking utilizes high-quality alternative labels, but suffers from coverage issues if relevant labels are not yet in Wikidata. Third, we plan to further develop our semantic graphs to cover other domains of question answering, such as non-factoid QA.

Acknowledgments. We thank the anonymous reviewers for their valuable comments and insights that helped us to improve upon the initial version of the paper.

This work has been supported by the German Research Foundation as part of the QA-EduInf project (grant GU 798/18-1 and grant RI 803/12-1). We gratefully acknowledge the support of NVIDIA Corporation with the donation of the Tesla K40 GPU used for this research.

References

1. Aghaebrahimian, A., Jurčíček, F.: Open-domain factoid question answering via knowledge graph search. In: Proceedings of 2016 NAACL Human-Computer Question Answering Workshop, pp. 22–28 (2016)
2. Bao, J., Duan, N., Yan, Z., Zhou, M., Zhao, T.: Constraint-based question answering with knowledge graph. In: Proceedings of the 26th International Conference on Computational Linguistics (COLING), pp. 2503–2514 (2016)
3. Berant, J., Chou, A., Frostig, R., Liang, P.: Semantic parsing on freebase from question-answer Pairs. In: Proceedings of the Conference on Empirical Methods in Natural Language Processing (EMNLP), pp. 1533–1544 (2013)
4. Berant, J., Liang, P.: Semantic parsing via paraphrasing. In: Proceedings of 52nd Annual Meeting of the Association for Computational Linguistics, pp. 1415–1425 (2014)
5. Bordes, A., Usunier, N., Chopra, S., Weston, J.: Large-scale Simple Question Answering with Memory Networks. arXiv preprint (2015)
6. Diefenbach, D., Singh, K., Maret, P.: WDAqua-core0: a question answering component for the research community. In: Proceedings of the 7th Open Challenge on Question Answering over Linked Data (QALD-7) at ESWC (2017)
7. Dong, L., Lapata, M.: Language to logical form with neural attention. In: Proceedings of the 54th Annual Meeting of the Association for Computational Linguistics, pp. 33–43 (2016)
8. Dong, L., Wei, F., Zhou, M., Xu, K.: Question answering over freebase with multi-column convolutional neural networks. In: Proceedings of the 53rd Annual Meeting of the Association for Computational Linguistics and the 7th International Joint Conference on Natural Language Processing, pp. 260–269 (2015)
9. Hakimov, S., Unger, C., Walter, S., Cimiano, P.: Applying semantic parsing to question answering over linked data: addressing the lexical gap. In: Biemann, C., Handschuh, S., Freitas, A., Meziane, F., Métais, E. (eds.) NLDB 2015. LNCS, vol. 9103, pp. 103–109. Springer, Cham (2015). doi:10.1007/978-3-319-19581-0_8
10. Huang, P., He, X., Gao, J., Deng, L., Acero, A., Heck, L.: Learning deep structured semantic models for web search using clickthrough data. In: The 22nd ACM International Conference on Information & Knowledge Management (CIKM), pp. 2333–2338 (2013)
11. Jain, S.: Question answering over knowledge base using factual memory networks. In: Proceedings of the 2016 Conference of the North American Chapter of the Association for Computational Linguistics: Human Language Technologies, pp. 109–115 (2016)
12. Kingma, D., Ba, J.: Adam: A Method for Stochastic Optimization. arXiv preprint (2014)
13. Lukovnikov, D., Fischer, A., Lehmann, J., Auer, S.: Neural network-based question answering over knowledge graphs on word and character level. In: Proceedings of the 26th International Conference on World Wide Web - WWW 2017, pp. 1211–1220 (2017)
14. Manning, C.D., Bauer, J., Finkel, J., Bethard, S.J., Surdeanu, M., McClosky, D.: The stanford CoreNLP natural language processing toolkit. In: Proceedings of 52nd Annual Meeting of the Association for Computational Linguistics, pp. 55–60 (2014)
15. Reddy, S., Lapata, M., Steedman, M.: Large-scale semantic parsing without question-answer Pairs. Trans. Assoc. Comput. Linguist. **2**, 377–392 (2014)

16. Reddy, S., Täckström, O., Collins, M., Kwiatkowski, T., Das, D., Steedman, M., Lapata, M.: Transforming dependency structures to logical forms for semantic parsing. Trans. Assoc. Comput. Linguist. **4**, 127–140 (2016)
17. Unger, C., Forascu, C., Lopez, V., Ngomo, A.-C.N., Cabrio, E., Cimiano, P., Walter, S.: Question answering over linked data (QALD-5). In: CEUR Workshop Proceedings, vol. 1391 (2015)
18. Van Der Maaten, L., Hinton, G.: Visualizing high-dimensional data using t-sne. J. Mach. Learn. Res. **9**, 2579–2605 (2008)
19. Vrandečić, D., Krötzsch, M.: Wikidata: a free collaborative knowledgebase. Commun. ACM **57**(10), 78–85 (2014)
20. Yao, X.: Lean question answering over freebase from scratch. In: Proceedings of the 2015 Conference of the North American Chapter of the Association for Computational Linguistics, pp. 66–70 (2015)
21. Yao, X., Van Durme, B.: Information extraction over structured data: question answering with freebase. In: Proceedings of the 52nd Annual Meeting of the Association for Computational Linguistics, pp. 956–966 (2014)
22. Yih, W.T., Chang, M., He, X., Gao, J.: Semantic parsing via staged query graph generation: question answering with knowledge base. In: Proceedings of the 53rd Annual Meeting of the Association for Computational Linguistics and the 7th International Joint Conference on Natural Language Processing, pp. 1321–1331 (2015)

WDAqua-core0: A Question Answering Component for the Research Community

Dennis Diefenbach[(✉)], Kamal Singh, and Pierre Maret

Université de Lyon, CNRS UMR 5516 Laboratoire Hubert Curien, 42023
Saint-Etienne, France
{dennis.diefenbach,kamal.singh,pierre.maret}@univ-st-etienne.fr

Abstract. We describe and present a new Question Answering (QA)
component that can be easily used by the QA research community.

It can be used to answer questions over DBpedia and Wikidata. The
language support over DBpedia is restricted to English, while it can be
used to answer questions in 4 different languages over Wikidata namely
English, French, German and Italian. Moreover it supports both full
natural language queries as well as keyword queries.

We describe the interfaces to access and reuse it and the services it can
be combined with. Moreover we show the evaluation results we achieved
on the QALD-7 benchmark.

Keywords: Question answering · Qanary · QALD

1 Introduction

Question answering (QA) is a very old research field in computer science. In
the last two decades, thanks to the development of the Semantic Web, a lot of
new structured data has become available on the web in the form of knowledge
bases (KBs). Nowadays, there are KBs about media, publications, geography,
life-science and more[1]. The idea behind a QA system over KBs is to find the
information, in a KB, requested by the user using natural language. This is
generally addressed by translating a natural question to a SPARQL query that
can be used to retrieve the desired information. We present here a QA component
to answer questions over DBpedia and Wikidata that can answer both full and
keyword natural language questions. It is integrated in the Qanary Ecosystem
[4] so that first, it can be easily reused by the research community and second,
it takes advantage of the services available in Qanary.

2 Related Work

In the context of QA, a large number of systems have been developed in the last
years. For example, more than twenty QA systems were evaluated against the

---

[1] http://lod-cloud.net.

© Springer International Publishing AG 2017
M. Dragoni et al. (Eds.): SemWebEval 2017, CCIS 769, pp. 84–89, 2017.
https://doi.org/10.1007/978-3-319-69146-6_8

QALD benchmark[2]. While many systems are querying DBpedia, we are only aware of one system querying wikidata, namely *Platypus*[3]. Moreover most of the works address full natural language questions while only few address keyword questions. One exception is *SINA*[7].

The fact that QA systems often reuse existing techniques lead to the idea of developing QA systems in a modular way. Four frameworks tried to achieve this goal: QALL-ME [5], openQA [6], the Open Knowledge Base and Question-Answering (OKBQA) challenge[4] and Qanary [1,4,8]. We integrated our QA component into the Qanary Ecosystem since it makes it easily reusable by the research community and offers a series of off-the-shelf services related to QA systems.

3 Description of WDAqua-core0

Our SPARQL creation algorithm uses a combinatorial approach based on the semantics encoded in the underlying KB. The full details will be disclosed in an upcoming publication as this is only a challenge submission. In the following we briefly describe the capabilities of WDAqua-core0. WDAqua-core0 can answer questions on both DBpedia and Wikidata. Note that the Wikidata dump[5] contains binary and non-binary relationships. An example of a non-binary relationships expressing that *the capital of Germany was Berlin from 1990* is expressed in two versions:

```
@prefix wd: <http://www.wikidata.org/entity/> .
@prefix p: <http://www.wikidata.org/prop/> .
@prefix ps: <http://www.wikidata.org/prop/statement/> .
@prefix wdt: <http://www.wikidata.org/prop/direct/> .
@prefix pq: <http://www.wikidata.org/prop/qualifier/> .
@prefix wds: <http://www.wikidata.org/entity/statement/> .

wd:Q183 rdfs:label "Germany"@en ;
wd:Q64 rdfs:label "Berlin"@en ;
wdt:P36 rdfs:label "capital"@en ;

#VERSION 1: reefied
wd:Q183 p:P36 wds:q183-7068B86F .
wds:q183-7068B86F a wikibase:Statement ,
    ps:P36 wd:Q64 ;
    pq:P580 "1990-10-03T00:00:00Z"^^xsd:dateTime .

#VERSION 2: non-reefied
wd:Q183 wdt:P36 wd:Q64 ;
```

---

[2] http://qald.sebastianwalter.org.

[3] https://askplatyp.us/?.

[4] http://www.okbqa.org/.

[5] https://dumps.wikimedia.org/wikidatawiki/entities/, https://www.mediawiki.org/wiki/Wikibase/Indexing/RDF_Dump_Format.

The first version uses properties with the namespaces p and ps while the second loses the temporal information and uses the namespace wdt. WDAqua-core0 is querying only the triples containing properties with namespace wdt. WDAqua-core0 can answer both keyword questions and questions in natural language. The complexity of the generated queries is limited to queries containing at most two triple patterns. The generated queries can be of type SELECT or ASK. The modifiers are limited to the COUNT operator. Thus, the questions with superlatives and comparatives can in general not be answered. Finally it supports English on DBpedia and 4 different language over Wikidata, namely English, French, German and Italian. The evaluation is shown in Sect. 5.

4 Integration in Qanary

Qanary is a framework to integrate QA components with the goal to make existing research in the QA field reusable. The QA component presented here is integrated into Qanary. A running version is registered into the Qanary service running under:

$$\text{http://www.wdaqua.eu/qanary}$$

In particular the component can be executed through RestFul interfaces. To run the service over a new question the RestFul interface under:

$$\text{http://www.wdaqua.eu/qanary/startquestionansweringwithtextquestion}$$

can be used. Besides the generated answer, the top-30 generated queries can also be retrieved.

The integration into Qanary allows the combination of WDAqua-core0 with the other components and services that are already integrated into Qanary. In particular it can be combined with a speech recognition component and a language detection component. Additionally it can be used together with a number of services that are constructed around Qanary. These include a reusable front-end called Trill [2]. A demo of Trill that in the back-end uses WDAqua-core0 can be found under www.wdaqua.eu/qa. Figure 1 shows a screen-shot of Trill. Moreover WDAqua-core0 can be used together with some interfaces for user-feedback that are integrated into Trill [3]. One such feedback-interface can be seen in Fig. 2. As a consequence WDAqua-core0 can be used by end-users and can for example be used to drive forward research in the domain of human-computer interaction. Finally Qanary has an interface that allows QA pipelines to be evaluated using Gerbil for QA[6]. This means that WDAqua-core0 can be evaluated by the research community at all time especially when new benchmarks arise.

---

[6] http://gerbil-qa.aksw.org.

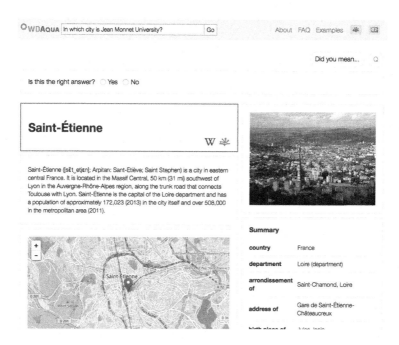

Fig. 1. Screenshot of Trill using in the back-end WDAqua-core0 for the question "In which city is Jean Monnet University?"

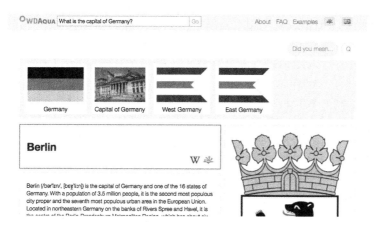

Fig. 2. Snapshot of the disambiguation interface for the question: "What is the capital of Germany?". By clicking on "Did you mean" several entities, the question might refereed to, are shown. These include the actual "Federal Republic of Germany" but also the "Capital of Germany" (as an entity), "West Germany", "East Germany", "Allied-Occupied Germany" and others. By clicking on the entity, the question is interpreted differently and a new answer is presented, e.g., if the user clicks on "West Germany", the answer "Bonn" is computed.

5   Evaluation over QALD-7

In this section we show the results of WDAqua-core0 over QALD-7 task 1 and task 4. We evaluate both over the keyword and the full-natural language questions.

Moreover, we extended the training set of task 4 and introduced a new type of multilingual QA benchmark. QALD-7 task 1 requires to answer questions in multiple languages using data contained in the English DBpedia. In particular taking the Italian DBpedia to answer the Italian questions of QALD-7 task 1 does not work in general. The fact that the Italian questions must be answered using the English dataset, forces the systems to use translations. Instead we translate the questions of the QALD-7 task 4 into French, German and Italian and try to answer them using Wikidata. This is fundamentally different since in Wikidata the knowledge is the same and only the labels change. In particular a translation is not required, one can answer the Italian questions using an Italian dataset.

The global (or macro) precision, recall and F-measure achieved over QALD-7 can be found in Table 1. Note that WDAqua-core0 does not use a machine learning algorithm so there is not a problem of over-fitting the dataset.

Table 1. The table shows the results of WDAqua-core0 over the QALD-7 training set.

Task	Type	Dataset	Language	Type	Precision	Recall	F-measure
1	train	DBpedia	en	full	0.49	0.54	0.51
1	train	DBpedia	en	keywords	0.37	0.41	0.39
1	test	DBpedia	en	full	0.30	0.30	0.30
1	test	DBpedia	en	keywords	0.08	0.10	0.09
4	train	Wikidata	en	full	0.32	0.32	0.32
4	train	Wikidata	en	keywords	0.28	0.28	0.28
4	test	Wikidata	en	full	0.39	0.40	0.40
4	test	Wikidata	en	keywords	0.32	0.33	0.33
4	train	Wikidata	fr	full	0.19	0.24	0.21
4	train	Wikidata	fr	keywords	0.33	0.37	0.34
4	train	Wikidata	de	full	0.220	0.23	0.22
4	train	Wikidata	de	keywords	0.30	0.35	0.33
4	train	Wikidata	it	full	0.16	0.18	0.17
4	train	Wikidata	it	keywords	0.19	0.21	0.20

6   Conclusion

We have presented a QA component integrated into the Qanary Ecosystem that can be easily reused by the QA community. In particular it can used to push forward research in directions like the integration of speech recognition systems with QA systems and the interaction with users.

We have evaluated the component against QALD-7 in multiple aspects. We have shown the performance over both DBpedia and Wikidata with respect to keyword and full-natural language queries. Moreover, we have introduced a new type of multilingual QA benchmark that does not require translation but where the questions and the KB are in the same language. We have shown our results over this new type of multilingual QA benchmark.

Acknowledgments. Parts of this work received funding from the European Union's Horizon 2020 research and innovation programme under the Marie Skodowska-Curie grant agreement No. 642795, project: Answering Questions using Web Data (WDAqua).

References

1. Both, A., Diefenbach, D., Singh, K., Shekarpour, S., Cherix, D., Lange, C.: Qanary a methodology for vocabulary-driven open question answering systems. In: ESWC 2016 (2016)
2. Diefenbach, D., Amjad, S., Both, A., Singh, K., Maret, P.: Trill: a reusable front-end for QA systems. In: ESWC P&D (2017)
3. Diefenbach, D., Hormozi, N., Amjad, S., Both, A.: Introducing feedback in qanary: How users can interact with QA systems. In: ESWC P&D (2017)
4. Diefenbach, D., Singh, K., Both, A., Cherix, D., Lange, C., Auer, S.: The qanary ecosystem: getting new insights by composing question answering pipelines. In: Cabot, J., Virgilio, R., Torlone, R. (eds.) ICWE 2017. LNCS, vol. 10360, pp. 171–189. Springer, Cham (2017). doi:10.1007/978-3-319-60131-1_10
5. Ferrández, Ó., Spurk, C., Kouylekov, M., Dornescu, I., et al.: The QALL-ME framework: A specifiable-domain multilingual Question Answering architecture. J. Web Sem. **9**(2) (2011). Elsevier
6. Marx, E., Usbeck, R., Ngonga Ngomo, A., Höffner, K., Lehmann, J., Auer, S.: Towards an open question answering architecture. In: SEMANTiCS (2014)
7. Shekarpour, S., Marx, E., Ngomo, A.C.N., Auer, S.: Sina: Semantic interpretation of user queries for question answering on interlinked data. Web Semant. Sci. Serv. Agents World Wide Web **30** (2015)
8. Singh, K., Both, A., Diefenbach, D., Shekarpour, S.: Towards a message-driven vocabulary for promoting the interoperability of question answering systems. In: ICSC 2016 (2016)

AMAL: Answering French Natural Language Questions Using DBpedia

Nikolay Radoev[1(✉)], Mathieu Tremblay[1], Michel Gagnon[1], and Amal Zouaq[2]

[1] Département de Génie Informatique et Génie Logiciel,
Polytechnique Montréal, Montreal, Canada
{nikolay.radoev,mathieu-4.tremblay,michel.gagnon}@polymtl.ca
[2] School of Electrical Engineering and Computer Science,
University of Ottawa, Ottawa, Canada
azouaq@uottawa.ca

Abstract. While SPARQL is a powerful way of accessing linked data, using natural language is more intuitive for most users. A few question answering systems already exist for English, but none focus specifically on French. Our system allows a user to query the DBpedia knowledge by asking questions in French separated in specific types, which are automatically translated into SPARQL queries. To our knowledge, this is the first French-based question answering system in the QALD competition.

Keywords: Question answering · Linked data · RDF · DBpedia · SPARQL

1 Introduction

A crucial aspect of making the Semantic Web relevant is providing typical Web users with an intuitive and powerful interface to access the growing amount of structured data. Among the datasets that are currently accessible to the public, there are general knowledge bases (KBs), such as DBpedia [1], which contains information extracted from Wikipedia, and many specialized knowledge bases that have curated domain-specific knowledge, such as Dailymed [2]. However, given their reliance on SPARQL [3], they are difficult to use for the average user. Moreover, querying these KBs without knowledge of their underlying structure is a very complex task. Developing an intuitive interface to allow natural language queries is a problem that has been explored in some of the previous QALD challenges [4].

QALD is a series of evaluation campaigns on question answering over linked data. Multiple systems have already been developed in the past years to solve this problem with up to 0.89 F-Score for English on Task 1 of the QALD competition [5] (Multilingual question answering over DBpedia). In this competition, most systems have been focused on English, primarily because of the amount of resources available in existing KBs [6] and its popularity worldwide. However, we have chosen to tackle questions asked in the French language using both the French and English chapters of DBpedia.

© Springer International Publishing AG 2017
M. Dragoni et al. (Eds.): SemWebEval 2017, CCIS 769, pp. 90–105, 2017.
https://doi.org/10.1007/978-3-319-69146-6_9

Because of French's specific language-dependent syntactic structures, it is not possible to directly reuse techniques that are available for the English language. For example, adjective placement in English precedes the noun while most of the time in French the adjective follows the noun it describes. Also, the syntactic rules to express verb tenses in French differ greatly from English: simple past in English may be expressed in French by a simple past form or a compound form (*talked* vs. *parla* and *a parlé*), there is no gerund form in French (*is talking* vs. *est en train de parler*), future tense is expressed by an auxiliary in English and by a suffix in French (*will talk* vs. *parlera*), etc.

Interpreting a question given in a natural language is a well-known but unsolved problem [7]. In general, it requires the extraction of a semantic representation that is the result of a multiple-phase approach. The question must be processed to extract keywords and terms that may represent some entities available in different KBs. Then, those keywords and terms must be mapped to resources, classes, and properties in the KBs. This is a complex task given the fact that (i) those keywords and terms might not exist as resources in the KBs and (ii) natural language syntax creates ambiguity that cannot be resolved without proper context. Questions such as *Who made Titanic?* (We are looking for the producer of the 1997 movie) are a good example of an ambiguity between the ship and the film and a need to infer the property http://dbpedia. org/ontology/producer from the verb *made*.

Some previous works on the problem used controlled natural language (CNL) [8] approaches to restrict grammar and syntax rules of the input question. Such approaches have the merit of reducing ambiguity and increasing the accuracy of the proposed answers. However, we consider the rigidity of an imposed grammar to be awkward for an average user and we do not impose any constraints on the questions given as input. This paper is a more detailed version of the one initially submitted [9] to the QALD-7 challenge. During the Open Challenge on Question Answering over Linked Data, the AMAL (Ask Me In Any Language) system distinguished itself for being the only system that focuses specifically on French.

For readability purposes, we have abbreviated certain URIs by using the prefixes detailed in Table 1. For example, http://dbpedia.org/ontology/spouse becomes *dbo:spouse*. The *dbo:* prefix is used to represent classes (aka concepts) from the DBpedia ontology while the *dbr:* prefix is used to identify resources from this KB. The *yago:* prefix refers to entities defined in the Yago knowledge base.

Table 1. DBpedia prefixes

dbo	http://dbpedia.org/ontology/
dbr	http://dbpedia.org/resource/
yago	http://dbpedia.org/class/yago/

2   Related Work

Previous work has already been done to answer natural language questions on a multilingual KB. We have focused our work on answering questions with the DBpedia [1] KB. This KB has been extracted from Wikipedia to represent general knowledge in multiple languages. Some systems also use Wikidata [10] to answer questions. The main difference between these two KBs is that DBpedia is automatically extracted from Wikipedia while Wikidata is manually created and supports the knowledge contained in Wikipedia.

QAKIS is a system that answers questions in English with the use of the French, English, Italian and German chapters of DBpedia [11]. It uses the WikiFramework repository (which contains relational patterns automatically· extracted from Wikipedia) [12] to solve the problem of finding lexicalizations of properties from the DBpedia ontology. Finding lexicalizations is a common problem in question answering since different wordings can be used to ask the same question.

WDAqua-core0 is another system presented at the 2017 ESWC conference that accepts many languages: French, English, German and Italian, with English being used with both DBpedia and Wikidata, whereas only Wikidata is used for other languages. The system is integrated in the Qanary Ecosystem [13] and uses some of its other features, most notably the speech recognition module. Just like QAKIS, WDAqua-core0 focuses on simple questions by translating natural language queries to SPARQL queries. Multilingual queries do not implement custom rules but rely on generic ones.

Our system differs from these because it is able to process questions in French and answer over the knowledge contained in DBpedia. QAKIS only processes English queries and while WDAqua-core0 can process questions in French, only knowledge from Wikidata is used and not DBpedia as specified in the QALD challenge. As far as we know, AMAL is the only system that specifically targets the French language.

3   System Overview

AMAL (Ask Me in Any Language) was developed using a modular approach to separate application logic in different systems. Each subsystem can be developed, modified and improved independently. With our system, users can ask questions in French that are analyzed and answered with information found in the English DBpedia, as specified in the description of Task 1 of the QALD challenge, for which this system was created.

Our system is the first version of a work in progress. It focuses on *Simple Questions*, which we define as questions that concern only one single entity and a single property of this entity, such as *Qui est le père de Barack Obama?* (Who is the father of Barack Obama?), where 'Barack Obama' is the entity and 'father of' is the property. We are still working on handling more complex questions involving multiple entity/property relations.

Our approach consists of a multiple-step pipeline: question type identification, entity extraction, property identification and question answering through a SPARQL query builder. The first step determines the type and possible subtype of the question using a *Question Type Analyzer*. Our system currently supports the following types: *Boolean, Date, Number and Resource*. Boolean questions are questions that can only be answered by *TRUE* or *FALSE*. Date questions refer to specific dates in a standard *YYYY-MM-DD* format. Number questions are questions that have answers in a numeric literal that is a value of one single property and not derived by using arithmetic manipulation. Finally, Resource questions are questions with answers given under the form of a DBpedia URI such as *dbr:Barack_Obama*. Two subtypes may be added to some of those main types: *List* and *Aggregation*. *List* questions, which can be subtype of both *Resource* and *Date* questions, are questions whose answer contains several elements. *Aggregation* questions include (i) questions that require an ascending or descending order such as looking for the most or least of something and (ii) questions that require counting, most often the number of elements that satisfy one or more conditions. *Date, Number and Resource* questions are the ones that may have *Aggregation* as a subtype. Our approach to classify questions is explained in Sect. 4.1.

Once the system knows the question type, the query is sent to specific *Question Solvers*. For instance, a question such as *Is Michelle the name of Barack Obama's wife?* will be sent to the boolean solver, and *When was Barack Obama born?* is handled by the date question solver. Aggregation Questions necessitate additional computation which is detailed in Sect. 5.3. Every question solver makes use of one or more submodules that function as *extractors*. There are two main extractors: an *entity extractor* and a *property extractor*, as shown on Fig. 1, which are used to identify the entities and properties in a given question. Specific question solvers require specific property extractor heuristics (more details are given in Sect. 4.3). Before the question is passed to its specific question solver, all question indicators, as seen in Tables 5, 6 and 7 are stripped from the question. In the case of Aggregation questions, the ordering indicators from Tables 2, 3 and 4 are also removed.

The AMAL system was created specifically for the 2017 QALD Task 1 challenge, where all answers are required to be extracted from the English version of DBpedia. Given that we focus on French questions, we need a way to translate the entities and/or properties found by our system. The *Translator* module handles the translation from French to English using Google Translate API for most queries. However, in our experience, Google Translate does sometimes give a different translation term than the one used by DBpedia. In those cases, the system uses custom translations to obtain the correct terms for the final query. As an example, the term *portée* (which means *span* when used as an attribute of a bridge or a road) is translated as *scope* instead of *span* and thus requires a specific rule to be correctly used. This is the last step before constructing the final SPARQL query. The additional overhead of French to English translation was added in order to conform with the competition's rules.

QUESTION

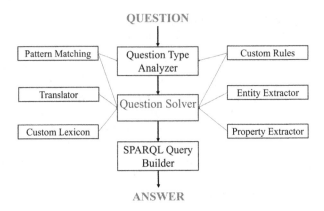

ANSWER

Fig. 1. System overview.

4   System Details

This section presents the details of the four main steps of our pipeline, which are question classification, entity extraction, property extraction and SPARQL query building. The description of each specific type of question solver is the subject of Sect. 5.

4.1   Question Classification

By analyzing the questions given in the 2016 and 2017 train datasets, we extracted various keywords and patterns that occur most often in a given question type. The extracted patterns rely both on lexical matching and positioning (start of a sentence or just their presence at any position in the question). For example, pronoun inversions such as *existe-t-il(elle)* (where the general form is *VERB-t-il(elle)*) appear only in close-ended (*Boolean*) questions.

Classification is made by matching the question string against the list of extracted patterns and if a match is found, the question gets assigned a specific type. The type matching is done by trying to match the questions to different types in the following order: Boolean, Number, Date and Resource. This order is due to the relative complexity of type matching with Boolean questions being easier to detect than Date questions, which might require additional work (see Sect. 5.1). In the current version of the system, multiple types are not supported and the first detected type is considered the only one. The system does however support subtypes, as explained later.

If no match is found after going through all extracted keywords and patterns, we assign a default value of *Resource* question type. The same method is then applied for the subtypes with a few additional tweaks. For *List* questions, we look for question words or verbs that indicate multiple answers, such as *Qui sont les* (Who are the). For *Aggregation*, we try to determine whether the answers require being sorted in a descending or an ascending order. This classification is made

Table 2. Lexical patterns for Aggregation count questions classification

Starts with	Example
Combien de	Combien de livres a écrit Isaac Asimov
Combien d'	Combien d'enfants a eu Barack Obama
Combien y	Combien y a-t-il de magasins Aldi

Table 3. Lexical patterns for Aggregation Descending order questions classification

Contains	Example
le plus vieux/la plus vieille	Donnez-moi le plus vieux président des États-Unis
le plus âgé/la plus âgée	Donnez-moi l'enfant le plus âgé de Barack Obama
le premier/la premiére	Donnez-moi le premier enfant de Barack Obama

Table 4. Lexical patterns for Aggregation Ascending order question classification

Contains	Example
le plus jeune/la plus jeune	Qui est le plus jeune enfant de Barack Obama
le plus haut/la plus haute	Quelle est la plus haute montagne d'Australie
le plus long/la plus longue	Quel est le pont le plus long
le dernier/la derniére	Qui est le dernier enfant de Barack Obama
le plus grand/la plus grande	Quelle est la plus grande montagne d'Australie

using a list of keywords and patterns extracted from the training datasets. Some patterns require the expression to occur only in the beginning of the sentence or anywhere in the sentence as highlighted by the first row of each table.

To determine whether a question is an aggregation question, we use the rules described in Tables 2, 3 and 4. The first column of each table describes the patterns used to classify a question in that category and the second one contains examples of questions for the category. We only check whether a sentence contains this expression to determine if it is an aggregation question.

In the same fashion, Tables 5, 6 and 7 shows our rules to classify a question as *Boolean*, *Date* or *Location*. For example, *Est-ce que les grenouilles sont des amphibiens?* is a boolean question because it starts with the expression *Est-ce que*, but *Quand est-ce que le Carey Price est né?* would not be considered a boolean question, since it does not start with the *est-ce que* pattern, but rather has it in the middle of the question. According to Table 5, it is classified as a *Date* question.

4.2 Entity Extraction

Once our system has identified the type of the question, it extracts the entity from the sentence. For example, with *Quand est-ce que Carey Price est né?*,

Table 5. Date question classification

Contains	Example
Quand	Quand a eu lieu l'Opération Overlord
Quelle est la date	Quelle est la date de naissance de Rachel Stevens
Donne moi la date	Donne moi la date de naissance de Rachel Stevens
À quelle date	À quelle date est née Rachel Stevens
Quelle est l'année/le mois	Quel est l'année de naissance de Rachel Stevens

Table 6. Boolean question classification

Starts with	Contains	Example
Est-ce que		Est-ce que le titanic est un bateau
Ont/Sont		Sont les grenouilles un type d'amphibien
Peut-on		Peut-on trouver des frèsques en Crète
	est-il/est-elle/sont-ils/sont-elles	Les grenouilles sont-elles des amphibiens

we extract *Carey Price* from the sentence. To do so, we use a syntactic parser to identify the noun groups and we then find the ones that correspond to an existing entity in DBpedia. Given that noun groups can contain many more elements than just the entity, such as adjectives or determinants, we start by taking the longest string in the question and generate all possible substrings by recursively tokenizing the output. All generated combinations are then ran against the DBpedia database to generate as many valid entities as possible. For example, *Who is the queen of England?* generates the following substrings after removing the question indicator (*Who is*): *queen of England, queen, England, queen of, of England.* Out of those, only the first 3 are kept as valid entities since *queen of* and *of England* are not DBpedia entities. Having multiple possible entities can help increase the chance of obtaining more accurate answers. Note that we translated the example "Who is the queen of England" in English for readability purposes but this step is actually done on a French question.

Since the question is in French, the *sameAs* link is used to find the corresponding URI in the English version of DBpedia, considering that only URIs from that DBpedia version are considered valid answers for the task 1 of QALD.

Table 7. Location question classification

Contains	Example
Où	Où est situé le Palais de Westminster
Dans quel pays/quelle ville	Dans quelle ville est la Tour Eiffel
Lieu	Quel est le lieu de naissance de Barack Obama
Endroit	À quel endroit se trouve la Tour Eiffel

If no such link is available, we use the previously described *Translator module* to generate a possible English entity. For the question: *Donne-moi les ingrédients d'un biscuit aux brisures de chocolat* (Give me the ingredients of a chocolat chip cookie), there is no French page *dbr:Cookie_aux_brisures_de_chocolat*, but using Google translate we are able to find *dbr:Chocolate_chip_cookie*, which is a valid DBpedia entity. Entities extracted this way however have a lower chance of being selected as the right entity given the uncertainty created by the translation.

To improve our results, we add variations of the identified nouns by applying modifications such as plurality indicators and capitalization. Every modification reduces the chance of the entity to be chosen as the main one. For example, the entity *queen* can also be manipulated in order to extract the entity *Queens*. However, since *Queens* requires 2 modifications (pluralization and capitalization), it is less likely to be selected than *queen*.

Once all the possible entities have been extracted, we use multiple criteria, such as the length of the entity's string, the number of modifications needed to extract it from DBpedia and whether we had to translate it, to determine their likelihood of being the right entity. The formula we currently use to combine these factors is the following, where e is the entity string:

$$length(e) - T \times \frac{length(e)}{2} + 3 \times U \times nsp(e)$$

where

$nsp(e)$ is the number of spaces in e

$T = 1$ if e has been translated, 0 otherwise

$U = 1$ if e has not been capitalized, 0 otherwise

According to this formula, the score is penalized by half the length of the entity string if we used a translation. Also, the more words it contains (whose value is obtained by counting the number of spaces) the higher the score will be, if it has not been capitalized. For example, *Dans quelle ville se trouve le Palais de Westminster?* we can extract *Palais de Westminster* and *Westminster*. However, *Westminster* by itself has a score of 11 while *Palais de Westminster* has a score of 27. Therefore, we consider it more likely that the correct entity is *Palais de Westminster*. For the question *Quels sont les ingrédients d'un biscuit au brisures de chocolat* we can extract *biscuit* and *chocolat*, with scores of 7 and 8. By translating the question to *What are the ingredients of a chocolate chip cookie?*, we can extract *chocolate chip cookie* with a score of 10.5, *chocolate chip* with a score of 10, *chocolate* with a score of 4.5 and *cookie* with a score of 3.

4.3   Property Extraction

Once the entity is extracted from the sentence, the property is found by removing the entity from the question and analyzing the remaining tokens to find all nouns or verbs. To find properties in the question, we use the DBpedia ontology and the RDF description of the selected entity in DBpedia. A property is found when we find a label that matches an URI in DBpedia and that is also present in the selected entity's description. As an example, in the question *Qui a créé Batman?* (Who created Batman?), the selected entity is *Batman* and we test if *dbr:Batman* contains a property that is also present in the phrase. In this specific case, the verb *created*, mapped to *dbo:creator* in our custom lexicon, does indeed appear in the description of *dbr:Batman*.

To facilitate the identification of the predicates to be used in the SPARQL query that corresponds to the question sentence, we built a lexicon by mapping a list of common DBpedia properties to French expressions, in addition to manually adding bindings that were not present in the French DBpedia. The creation of such a lexicon was necessary given the lack of similar resources in the French DBpedia or other external solutions that focus on French. For example, *dbo:spouse* and http://fr.dbpedia.org/property/conjoint are both mapped to *conjoint* (spouse), *épouse* (wife), *femme* (wife) and *mari* (husband). With such bindings, the system is able to take into account the various ways of expressing the property in French: *Qui est la conjointe/l'épouse/la femme de Barack Obama?* (Who is the spouse/wife/ wife of Barack Obama?).

4.4 SPARQL Query Builder

The last step is building and sending SPARQL queries to DBpedia. The system uses the standard DBpedia endpoint: http://dbpedia.org. For now, AMAL supports basic SPARQL queries such as *ASK* and *SELECT*. In addition, *COUNT* and *ORDER BY* queries are supported for *Aggregation* questions. Queries exist as basic templates in which specific values are injected in the form of RDF triples. For example, in *Qui est l'épouse de Barack Obama?*, once we have extracted *dbr:Barack_Obama* and *dbo:spouse*, our system will build the following query: *SELECT DISTINCT ?uri WHERE {dbr:Barack_Obama dbo:spouse ?uri}* and send it to DBpedia, which will return the answer. The library used is available on the following GitHub link: https://www.github.com/Mathos1432/dotNetSPARQL.

5   Simple Question Analysis

As previously mentioned, in our current implementation, we deal mostly with simple questions, limited to at most one entity and one property. In the case of boolean questions, we also process *Entity - Entity* relations, for example *Est-ce que la femme de Barack Obama est Michelle Obama?* (Is Michelle Obama the wife of Barack Obama?).

We start by identifying the type of the question, as explained in Sect. 4.1. Once this is done, we extract properties and entities. From there, we proceed to answering the question. To do so, we start with the most likely entity that was extracted, using the likelihood formula defined in Sect. 4.2, and execute SPARQL queries to determine whether an element in the rest of the question is a property of the entity. If the property exists, the object of the property is usually our answer.

When the entity does not have any of the extracted properties, we attempt disambiguation on the entity by following the Wikipage disambiguation link, if such a link is available. This link provides a way to find other entities that can be a possible match. For example, in *Who is the producer of Titanic?*, our system would extract the entity *Titanic*. Using the disambiguates link in *Titanic*, we can find *Titanic(film)*, *Titanic(album)* and other possible entities. The property *dbo:producer* can be found in the *Titanic(film)* resource description, thus allowing the system to select the correct entity.

5.1   Date and Location Questions

Some queries require that our system find dates or locations for events such as the birth of a person or the place of a war. For most of these questions, the label for the property is directly present in the query string. For example, in the question *Quelle est la date de naissance de Rachel Stevens* (What is Rachel Stevens birth date?), the property is given explicitly in the question (birth date). There are also questions where this is not the case and the property must be inferred, for example in *Quand Rachel Stevens est-elle née?* (When was Rachel Stevens born?). Our solution for this problem is using a lexicon that maps commonly used words or expressions such as *"année de naissance"* (year of birth) or *"est né(e)"* (was born) to the *dbo:dateOfBirth* property. Other types of dates, such as death date and end of career date, can also easily be handled in the same way.

Date questions require some additional work when comparison and ordering are involved and this is explained with more details in Sect. 5.3. Tables 8 and 9 show the lexicons we built to handle date and location questions.

Table 8. Lexicon of date properties

Keywords	Possible properties
mort	death date
année de naissance	birth year
naissance, né	birth date
dissolue	dissolution date
commencé	active years start date, active years start year
terminaison	active years end date, active years end year
terminé	completion date
fondé	founding year
indépendance	founding date

Table 9. Lexicon of location properties

Keywords	Possible properties
mort	death place
naissance	birth place
commence	route start, source country
enterré	resting place, place of burial
vin	wine produced
fondé	founding year
vit	residence

Location Questions. An additional step is done for location questions. Once our system found answers, it filters them to make sure we only consider the relevant ones. For example, the question *Dans quelle ville est situé le Palais de Westminster?* (In which city is the Palace of Westminster located?), if we query *dbr:Palace_of_Westminster* for its *dbo:location*, we will get *dbr:City_of_Westminster, dbr:Greater_London, dbr:United_Kingdom* and *dbr:England.* Since we are only looking for the city, we need to filter the answers to those that are of the right type using the property *rdfs:type*, which in this case is the type *yago:City108524735*, thus eliminating all entities except *dbr:City_of_Westminster.*

5.2   Boolean Questions

The first step in the processing of boolean questions is to determine whether the question involves other entities, as in *Is Michelle Obama the wife of Barack Obama?* (note that in this example *Barack Obama* is identified as the concerned entity). If it is the case, the program verifies whether a relation exists between the entities and if it is the right type (in this case, a relation of type *spouse*). When the question does not involve other entities, it is about whether a property exists for a specific entity. We can consider the example *Existe-t-il un jeu vidéo appelé Battle Chess?* (Is there a video game called Battle Chess?). In this case, we simply need to find all entities with a label *Battle Chess* and find out if one of them is a video game. Here again, we rely on the presence of *Existe* and look for a *rdf:type* relation using a mapping available in our lexicon. The last type of question supported by our system is whether an entity is of a specific type. For instance, *Are tree frogs a type of Amphibian?*. In this case, we extract the entity from the sentence (tree frog) and verify if it has the right type (Amphibian).

5.3   Aggregation Questions

Aggregation questions are divided into two different categories. The first category contains questions that are looking for a numeric answer, such as *Combien de langues sont parlées en Colombie?* (How many languages are spoken in

Colombia?). This type of question is characterized by the presence of numeric indicators, such as: *How many, All, Number of*, etc. In order to determine the entity subject to enumeration, all possible entities and properties are extracted from the question, with entities/properties immediately after the numeric indicator being considered as more likely to be the main subject. In the case of specific numeric indicators, additional work must be done since the counting is done based on the sentence's verb and not a particular entity. For example, *Combien de fois s'est mariée Jane Fonda?* (How often did Jane Fonda get married?), the enumeration is done on the number of spouses (*dbo:spouse*) and not Jane Fonda.

The second category contains questions that involve ordering result sets in a specific order. Key indicators of this question type are superlative adjectives present in the query. Expressions such as *le plus grand* (the most), *le plus gros* (the biggest), etc., are used to decide the sorting order. The queries are parsed before the *Question Solver* module and if a superlative adjective is found, it is compared to prebuilt custom lists to determine if an ascending or descending order is required. Adjectives such as *le plus grand, le plus gros (biggest, largest)* are marked as *Descending* indicators while *le plus petit, le plus facile (smallest, easiest)* are *Ascending* ones. The only exception to this rule are queries about dates or age (a type of implicit date) given that asking for *le plus jeune (the youngest)* of something is equivalent to looking for the highest date when ordering potential results. As an example for the following two dates: *1970-24-04* and *1969-23-11*, the first date would be considered as *youngest* even if *1970* is bigger than *1969*.

Once the ordering indicator is found and classified, it is removed from the query. This allows for an easier analysis of the question and reduces any possible ambiguity. As an example, in *Quelle est la plus haute montagne d'Australie?* (What is the highest mountain in Australia?), our system identifies *la plus haute* (the highest) as a *Descending* order indicator. After the question indicator *Quelle est (What is)* and the order indicator (the highest) are removed from the query, the two entities *dbo:Mountain* and *dbo:Australia* can be extracted. The system can then try to find a relationship between them, specifically looking for all entities of type *dbo:Mountain* that have a property with a value of *dbo:Australia* and then sorting them by the property *dbo:elevation* in a *Descending* order.

The relation between adjectives and entities can be somewhat ambiguous in most KBs. In fact, the adjective *la plus haute*(the tallest) can be applied to both a *Person* and *Object* (mountain for example). However, in the first case, the KBs property would be *height*, while in the second, we need to be looking for *elevation*. Our system offers a limited context awareness by using manually created *(adjective:entity),property* pairs. In the previously given example, the system is looking for *dbo:elevation*, since the pair {(*"la plus haute": dbo:NaturalPlace), dbo:elevation*} and {(*"la plus haute": dbo:Person), dbo:height*} exist in the predefined custom pairs and *dbo:mountain* is of type *dbo:NaturalPlace*.

The system does not currently support superlative ordering with multiple answers. Queries asking for the *N most* or *least* Entities, such as *Donnez-moi les 5 premiers présidents des États-Unis* (Give me the five first presidents of the United-States) will currently only return the first result.

5.4   List Questions

List questions are those that can but do not necessarily require an answer containing multiple entries. They can be seen as a subset of aggregation questions, but without additional sorting or counting over the result set. For example, *Who are the founders of DBpedia?* is a list question that returns multiple resources. Given that the type of results can vary, list questions are first analyzed based on their type.

Some types, such as *Boolean* and *Number*, are guaranteed to have only a single possible answer and all additional answers beyond the first ones that are returned by the SPARQL query are ignored. *Resource* and *Dates* types can return multiple possible answers, so all results returned by the SPARQL queries are added to the final result set. In the current version of the system, there is no imposed limit to the number of elements returned given the inherent difficulty of predicting the expected size of the answer for some questions.

Some questions can ask for a single entity but actually have multiple answers, such as *Qui était le successeur de John F. Kennedy?* (Who **was the successor** of John F. Kennedy?), which has 3 different answers in DBpedia (Lyndon B. Johnson, Benjamin A. Smith III and Tip O'Neill) despite us looking for a single entity. For this reason, our system does not restrict the result set to a single answer. An exception is made for *Aggregation* questions, since the results of a *Aggregation-Counting* question is processed as a *Number* response using a single literal as an answer. For example, *Combien de livres a écrit Isaac Asimov?* (How many books did Isaac Asimov write) is supported. But we do not currently support *Aggregation-Ordering* queries that return more than a single result, such as *Donnez-moi les 5 premiers présidents des États-Unis* (Give me the five first presidents of the United-States).

6   Evaluation

To evaluate the efficiency of our system, we used the train and test datasets for Task 1 of QALD-7. Table 10 shows, for each dataset, the number of questions and precision, recall and F-Score. The train dataset was evaluated using GERBIL-QA [14], while the test set was manually compared to the provided answers provided for the competition. Any results provided by our system that did not exactly match the provided answers (containing more or less elements for multiple element answers) were considered as failed.

Table 10. Results on the QALD 7 datasets

Dataset	Number of questions	Precision	Recall	F-Score
QALD-7 train dataset	214	0.9708	0.6967	0.8112
QALD-7 test dataset	50	0.7046	0.88	0.7825

7   Future Work

For further improvement of the AMAL system, we have considered many possible approaches. First, we would like to be able to improve our disambiguation algorithm. In its current version, AMAL looks through all possible disambiguation resources, whenever available, and attempts to find the property it extracted in the description of each resource. However, this can quickly fail when many entities have the property, especially when looking for more generic properties such as a birth date or location.

To improve our entity extraction, we plan to look at whether possible entities contain properties from the question. This would allow us to be more precise in our extraction and limit the amount of disambiguation we have to do, or at least help with it. For example, for the sentences *What is Battleship's budget?* and *Which platforms support Battleship?*, we can extract both the Battleship movie and the Battleship video game. However, in the first case, with the budget property we rank the movie with a higher likeliness of being the right entity. In the case of the video game, we can use the platform property to determine that we are looking for a video game rather than a movie.

There is also the possibility to add more lexicalization of DBpedia properties to our system as it only covers the ones we thought were most important for the challenge. To do so, an automated approach would help improve our results while reducing the amount of time it takes to build such a dataset. The WikiFramework could be used, or the lexicalization dataset from DBpedia-spotlight.

We also plan to process more complex queries involving more than one entity and property, starting by processing queries with multiple properties. This will be achieved by chaining requests until no properties are left to answer in the sentence. For example, *Quelle est la date de naissance du créateur de Dracula?* (What is the birthdate of the creator of Dracula), we can extract *dbo:birthDate* and *dbo:creator* from the sentence. Once this is done, we use our question solvers to link the properties in reverse order. In this case, we query DBpedia for Dracula's creator and then for this answer's birth date. This approach does not cover more complex cases but it is a realistic improvement goal.

Given that our system was built by using separate modules, we hope to be able to extend the custom logic modules that are directly affected by the French languages particularities allowing for an implementation of the system in other languages, mainly English, thus rendering it a truly multilingual Question-Answering system.

The use of multiple KBs to answer questions is another improvement that could be made in the future, which would allow the use of a greater body of knowledge to answer questions. The use of Wikidata would improve our ability to answer general knowledge questions.

8   Conclusion

The AMAL system is one of systems in the QALD competition that supports the French language. However, unlike other systems, AMAL focuses specifically on French questions and is thus able to obtain better and more accurate results. In this paper we have outlined the main functionality of the system and its modular, custom rule driven approach to Question Answering over Linked Data. Possible improvements are also outlined as both an analysis of the system current limitations and a source of any future work done on our system.

In its current version, DBpedia is offered in more than 30 languages including English, French, Spanish and many others. While the English version is by far the most complete and rich knowledge base, other versions can offer a substantial amount of information and knowledge. We hope that our work on a system that focuses on a language other than English can benefit other works that focus on more than the single dominating language that is English in the present time.

References

1. Auer, S., Bizer, C., Kobilarov, G., Lehmann, J., Cyganiak, R., Ives, Z.: DBpedia: a nucleus for a web of open data. In: Aberer, K., Choi, K.-S., Noy, N., Allemang, D., Lee, K.-I., Nixon, L., Golbeck, J., Mika, P., Maynard, D., Mizoguchi, R., Schreiber, G., Cudré-Mauroux, P. (eds.) ASWC/ISWC -2007. LNCS, vol. 4825, pp. 722–735. Springer, Heidelberg (2007). doi:10.1007/978-3-540-76298-0_52
2. National Institutes of Health et al. Daily med (2014)
3. https://www.w3.org/TR/sparql11-overview/
4. Lopez, V., Unger, C., Cimiano, P., Motta, E.: Evaluating question answering over linked data. Web Semant. Sci. Serv. Agents World Wide Web **21**, 3–13 (2013)
5. Unger, C., Ngomo, A.-C.N., Cabrio, E.: 6th open challenge on question answering over linked data (QALD-6). In: Sack, H., Dietze, S., Tordai, A., Lange, C. (eds.) SemWebEval 2016. CCIS, vol. 641, pp. 171–177. Springer, Cham (2016). doi:10.1007/978-3-319-46565-4_13
6. Lehmann, J., Isele, R., Jakob, M., Jentzsch, A., Kontokostas, D., Mendes, P.N., Hellmann, S., Morsey, M., van Kleef, P., Auer, S., Bizer, C.: Semantic Web. Dbpedia - a large-scale, multilingual knowledge base extracted from wikipedia, vol. 6 pp. 167–195 (2015)
7. Gupta, P.: A survey of text question answering techniques. Int. J. Comput. Appl. (2012)
8. Mazzeio, G.: Answering controlled natural language questions on RDF knowledge bases (2016). https://openproceedings.org/2016/conf/edbt/paper-259.pdf
9. Radoev, N., Tremblay, M., Gagnon, M., Zouaq, A.: Answering natural language questions on RDF knowledge base in French. In: 7th open challenge in Question Answering over Linked Data (QALD-7), Portoroz, Slovenia, May 2017

10. Vrandečić, D., Krötzsch, M.: Wikidata: a free collaborative knowledgebase. Commun. ACM **57**(10), 78–85 (2014)
11. Cabrio, E., Cojan, J., Gandon, F., Hallili, A.: Querying multilingual DBpedia with QAKiS. In: Cimiano, P., Fernández, M., Lopez, V., Schlobach, S., Völker, J. (eds.) ESWC 2013. LNCS, vol. 7955, pp. 194–198. Springer, Heidelberg (2013). doi:10. 1007/978-3-642-41242-4_23
12. Cabrio, E., Cojan, J., Magnini, B., Gandon, F., Lavelli, A.: QAKiS @ QALD-2. In: 2nd Open challenge in Question Answering over Linked Data (QALD-2), Heraklion, Greece, May 2012
13. Both, A., Diefenbach, D., Singh, K., Shekarpour, S., Cherix, D., Lange, C.: Qanary – a methodology for vocabulary-driven open question answering systems. In: Sack, H., Blomqvist, E., d'Aquin, M., Ghidini, C., Ponzetto, S.P., Lange, C. (eds.) ESWC 2016. LNCS, vol. 9678, pp. 625–641. Springer, Cham (2016). doi:10.1007/978-3-319-34129-3_38
14. Usbecka, R., ödera, M., Hoffmanna, M., Conradsa, F., Huthmanna, J., Ngonga-Ngomoa, A.-C., Demmlera, C., Ungerb, C.: Benchmarking question answering systems

Semantic Sentiment Analysis

Semantic Sentiment Analysis Challenge at ESWC2017

Diego Reforgiato Recupero[1(✉)], Erik Cambria[2], and Emanuele Di Rosa[3]

[1] Department of Mathematics and Computer Science, University of Cagliari, Cagliari, Italy
diego.reforgiato@unica.it
[2] Nanyang Technological University, Singapore, Singapore
cambria@ntu.edu.sg
[3] FINSA s.p.a, Genova, Italy
emanuele.dirosa@finsa.it

Abstract. Sentiment Analysis is a widely studied research field in both research and industry, and there are different approaches for addressing sentiment analysis related tasks. Sentiment Analysis engines implement approaches spanning from lexicon-based techniques, to machine learning, or involving syntactical rules analysis. Such systems are already evaluated in international research challenges. However, Semantic Sentiment Analysis approaches, which take into account or rely also on large semantic knowledge bases and implement Semantic Web best practices, are not under specific experimental evaluation and comparison by other international challenges. Such approaches may potentially deliver higher performance, since they are also able to analyze the implicit, semantics features associated with natural language concepts. In this paper, we present the fourth edition of the Semantic Sentiment Analysis Challenge, in which systems implementing or relying on semantic features are evaluated in a competition involving large test sets, and on different sentiment tasks. Systems merely based on syntax/word-count or just lexicon-based approaches have been excluded by the evaluation. Then, we present the results of the evaluation for each task and show the winner of *the most innovative approach* award, that combines several knowledge bases for addressing the sentiment analysis task.

1 Introduction

The development of Web 2.0 has given users important tools and opportunities to create, participate and populate blogs, review sites, web forums, social networks and online discussions. Tracking emotions and opinions on certain subjects allows identifying users' expectations, feelings, needs, reactions against particular events, political view towards certain ideas, etc. Therefore, mining, extracting and understanding opinion data from text that reside in online discussions is currently a hot topic for the research community and a key asset for industry.

The produced discussion spanned a wide range of domains and different areas such as commerce, tourism, education, health, etc. Moreover, this comes back and feeds the Web 2.0 itself thus bringing to an exponential expansion.

© Springer International Publishing AG 2017
M. Dragoni et al. (Eds.): SemWebEval 2017, CCIS 769, pp. 109–123, 2017.
https://doi.org/10.1007/978-3-319-69146-6_10

This explosion of activities and data brought to several opportunities that can be exploited in both research and industrial world. One of them concerns the mining and detection of users' opinions which started back in 2003 (with the classical problem of polarity detection) and several variations have been proposed. Therefore, today there are still open challenges that have raised interest within the scientific community where new hybrid approaches are being proposed that, making use of new lexical resources, natural language processing techniques and semantic web best practices, bring substantial benefits.

Computer World[1] estimates that 70%–80% of all digital data consists of unstructured content, much of which is locked away across a variety of different data stores, locations and formats. Besides, accurately analyzing the text in an understandable manner is still far from being solved as this is extremely difficult. In fact, mining, detecting and assessing opinions and sentiments from natural language involves a deep (lexical, syntactic, semantic) understanding of most of the explicit and implicit, regular and irregular rules proper of a language.

Existing approaches are mainly focused on the identification of parts of the text where opinions and sentiments can be explicitly expressed such as polarity terms, expressions, statements that express emotions. They usually adopt purely syntactical approaches and are heavily dependent on the source language and the domain of the input text. It follows that they miss many language patterns where opinions can be expressed because this would involve a deep analysis of the semantics of a sentence. Today, several tools exist that can help understanding the semantics of a sentence. This offers an exciting research opportunity and challenge to the Semantic Web community as well. For example, sentic computing is a multi-disciplinary approach to natural language processing and understanding at the crossroads between affective computing, information extraction, and common-sense reasoning, which exploits both computer and human sciences to better interpret and process social information on the Web.

Therefore, the Semantic Sentiment Analysis Challenge looks for systems that can transform unstructured textual information to structured machine processable data in any domain by using recent advances in natural language processing, sentiment analysis and semantic web.

By relying on large semantic knowledge bases, Semantic Web best practices and techniques, and new lexical resources, semantic sentiment analysis steps away from blind use of keywords, simple statistical analysis based on syntactical rules, but rather relies on the implicit, semantics features associated with natural language concepts. Unlike purely syntactical techniques, semantic sentiment analysis approaches are able to detect sentiments that are implicitly expressed within the text, topics referred by those sentiments and are able to obtain higher performances than pure statistical methods.

The fourth edition of the Semantic Sentiment Analysis Challenge[2] followed the success, experience and best practices of the first three. It provided further

---

[1] Computer World, 25 October 2004, Vol. 38, NO 43.

[2] https://github.com/diegoref/SSAC2017.

stimulus and motivations for research within the Semantic Sentiment Analysis area.

The fourth edition of the challenge focused on further development of novel approaches for semantic sentiment analysis. Participants had to design a concept-level opinion-mining engine that exploited Linked Data and Semantic Web ontologies, such as DBPedia[3].

The authors of the competing systems showed how they employed semantics to obtain valuable information that would not be caught with traditional sentiment analysis methods. Accepted systems were based on natural language and statistical approaches with an embedded semantics module, in the core approach. As happened within the first three editions of the challenge [14–16], a few systems merely based on syntax/word-count were excluded.

The fourth challenge benefited from a Google Group that we created and named Semantic Sentiment Analysis Intiative[4] and that we opened before the Challenge proposal. Currently, the group consists of more than 200 participants and we leverage that to disseminate and promote our initiatives related to the Sentiment Analysis domain. Moreover, the fourth edition of the challenge could also benefit from a Workshop[5] we chaired at ESWC 2017 related to the same topics. Challenge had therefore an additional strength provided by the mutual support between the two events. Challenge systems were in fact invited to a Workshop dedicated session for discussing open issues and research directions showing the last technological advancements. This dual action stimulated and encouraged participants to present their work at the two events.

The remainder of the chapter is organized as follows. Section 2 discusses background work related to semantic sentiment analysis. Section 3 lists and details the tasks we have proposed in the fourth edition of the challenge as well as the annotated datasets we have used for the training, testing and evaluation phase. Section 4 shows the systems submitted by the challengers and their results are showed in Sect. 5. Finally, conclusions, considerations and our plans for the next edition of the challenge are drawn in Sect. 6.

2   Related Work

After the successes of the 2014, 2015 and 2016 editions [14–16], the ESWC conference[6] included again a challenge call with a dedicated session. The Semantic Sentiment Analysis challenge has been proposed and accepted for the fourth time on a row in the 2017 ESWC program.

The 2014, 2015 and 2016 editions of the ESWC challenges have been published in books [5,13,19] where each challenge, its tasks, evaluation process have been introduced and each system participating to each challenge has been

---

[3] http://dbpedia.org.

[4] Publicly accessible at https://groups.google.com/forum/#!forum/semantic-senti ment-analysis.

[5] http://www.maurodragoni.com/research/opinionmining/events/.

[6] http://2016.eswc-conferences.org/.

described, detailed and results and comparisons have been shown. The Semantic Sentiment Analysis challenge has been included in the three volumes above [14–16]. The 2014 edition of the challenge was also the first edition in parallel with a workshop at ESWC of the same domain that hosted around 20 participants [7]. The 2016 and 2017 editions of the challenge repeated the success of the dual events of the 2014 edition and run in parallel with the Semantic Sentiment Analysis workshop whose proceedings are in the process of publication.

Besides the Semantic Sentiment Analysis challenge described in this chapter and its previous editions, there are a few number of relevant events and challenges that is worth to mention.

SemEval (Semantic Evaluation)[7] consists of a series of evaluations workshops of computational semantic analysis systems. It is now in its eleventh edition[8] and it has been collocated with the 55th annual meeting of the Association for Computational Linguistics (ACL)[9]. Since 2007 the workshop has covered the sentiment analysis topic. During the last edition, SemEval2017 included five tasks for the sentiment analysis track:

- Sentiment Analysis in Twitter. It was subdivided in five subtasks related to message polarity classification, topic-based message polarity classification and tweet quantification. The used languages were English and Arabic and the challenge organizers encouraged to use profile information provided in Twitter such as demographics (e.g. age, location) to analyze its impact on improving sentiment analysis.
- Fine-Grained Sentiment Analysis on Financial Microblogs and News. Divided in two tracks: one related to StockTwits messages consisting of microblog messages focusing on stock market events and assessments from investors and traders, exchanged via the StockTwits microblogging platform and the other related to Twitter messages consisting of tweets about stock market discussion within the Twitter platform. The problem was to predict the sentiment score for each of the companies/stocks mentioned where the sentiment values need to be floating point values in the range of −1 (very negative/bearish) to 1 (very positive/bullish), with 0 designating neutral sentiment.
- #HashtagWars: Learning a Sense of Humor. The goal of this task was to learn to characterize the sense of humor represented in a given show. Given a set of hashtags, the goal was to predict which tweets the show will find funnier within each hashtag. The degree of humor in a given tweet is determined by the labels provided by the show.
- Detection and Interpretation of English Puns. Puns are a class of language constructs in which lexical-semantic ambiguity is a deliberate effect of the communication act. That is, the speaker or writer intends for a certain word or other lexical item to be interpreted as simultaneously carrying two or more separate meanings. The task is divided into three subtasks where puns must be detected, localized and interpreted.

---

[7] https://en.wikipedia.org/wiki/SemEval.

[8] http://alt.qcri.org/semeval2017/.

[9] http://acl2017.org/.

– RumourEval: Determining rumour veracity and support for rumours. This task aimed to identify and handle rumours and reactions to them, in text.

Works such as [3, 6, 17] represent strong contributions within the domain of semantic sentiment analysis. In those the authors exploited unsupervised techniques to analyse the semantics of a given sentence providing information such as the opinion holder, the topic and the opinion being expressed by the holder to the topic.

Last but not least, authors in [2] provided a feasible research platform for the development of practical solutions for sentiment analysis to be beneficial for our society, business and future research as well.

3   Tasks, Datasets and Evaluation Measures

The fourth edition of the Semantic Sentiment Analysis challenge included six tasks: Polarity Detection, Polarity Detection in presence of metaphorical language, Aspect-Based Sentiment Analysis, Semantic Sentiment Retrieval, Frame Entities Identification, Subjectivity and Objectivity detection. One more task was represented by the Most Innovative Approach. Participants had to submit an abstract of no more than 200 words and a 4 pages paper including the details of their systems, why it is innovative, which features or functions it provides, which design choices were made, what lessons were learnt, which tasks it addressed and how the semantics was employed. Industrial tools with non disclosure restrictions were also allowed to participate, and in this case they were asked to:

– explain even at a higher level their approach and engine macro-components, why it is innovative, and how the semantics is involved;
– provide free access (even limited) for research purposes to their engine, especially to make repeatable the challenge results or other experiments possibly included in their paper.

As the challenge focused on the introduction, presentation, development and discussion of novel approaches to semantic sentiment analysis, participants had to design a semantic opinion-mining engine that exploited Semantic Web knowledge bases, e.g., ontologies, DBpedia, etc., to perform multi-domain sentiment analysis. Systems not including semantics have been rejected whereas the others had to provide a full description of their system, web access or a link where the system could be downloaded together with a short set of instructions. Moreover, accepted systems had to be either accessible via web or downloadable or anyway a RESTful API had to be provided to run the challenge test-set. If an application was not publicly accessible, password had to be provided for reviewers. A short set of instructions on how to use the application or the RESTFul API had to be provided as well.

Following we will describe each task and, in particular, will detail datasets and evaluation methodologies we have provided for tasks 1 and 3, those targeted by the submitted systems.

3.1   Task 1: Polarity Detection

The proposed semantic opinion-mining engines were assessed according to precision, recall and F-measure of the detected polarity values (positive OR negative) for each review of the evaluation dataset. As an example, for the tweet *GOOD LOOKING KICKS IF YOUR KICKIN IT OLD SCHOOL LIKE ME. AND COMFORTABLE. AND RELATIVELY CHEAP. I'LL ALWAYS KEEP A PAIR OF STAN SMITH'S AROUND FOR WEEKENDS*, the correct answer that a sentiment analysis system needed to give was *positive* and therefore it had to write *positive* between the *<polarity>*, *</polarity>* tags of the output. Figure 1 shows an example of the output schema for task1.

```xml
<?xml version="1.0" encoding="UTF-8" standalone="yes"?>
<Sentences>
    <sentence id="apparel_0">
        <text>
            GOOD LOOKING KICKS IF YOUR KICKIN IT OLD SCHOOL LIKE ME. AND COMFORTABLE.
            AND RELATIVELY CHEAP. I'LL ALWAYS KEEP A PAIR OF STAN SMITH'S
            AROUND FOR WEEKENDS
        </text>
        <polarity>
        positive
        </polarity>
    </sentence>
    <sentence id="apparel_1">
        <text>
            These sunglasses are all right. They were a little crooked, but still cool..
        </text>
        <polarity>
        positive
        </polarity>
    </sentence>
</Sentences>
```

Fig. 1. Task 1 output example. Input is the same without the polarity tag.

This task was pretty straightforward to evaluate. A precision/recall analysis was implemented to compute the accuracy of the output for this task. A true positive (tp) was defined when a sentence was correctly classified as positive. On the other hand, a false positive (fp) is a positive sentence which was classified as negative. Then, a true negative (tn) is detected when a negative sentence was correctly identified as such. Finally, a false negative (fn) happens when a negative sentence was erroneously classified as positive. With the above definitions, we defined the precision as

$$precision = \frac{tp}{tp + fp}$$

the recall as

$$recall = \frac{tp}{tp + fn}$$

the F1 measure as

$$F1 = \frac{2 \times precision \times recall}{precision + recall}$$

and the accuracy as

$$accuracy = \frac{tp + tn}{tp + fp + fn + tn}$$

As training, development and test sets, we used one million of reviews collected from the Amazon web site and split in 20 different categories: *Amazon Instant Video, Automotive, Baby, Beauty, Books, Clothing Accessories, Electronics, Health, Home Kitchen, Movies TV, Music, Office Products, Patio, Pet Supplies, Shoes, Software, Sports Outdoors, Tools Home Improvement, Toys Games, and Video Games*. The classification of each review (positive or negative) has been done according to the guidelines used for the construction of the Blitzer dataset [9]. Participants evaluated their system by applying a cross-fold validation over the dataset where each fold is clearly delimited. The script to compute Precision, Recall, and F-Measure and the confusion matrix has been provided to participants through the website of the challenge.

3.2   Task 2: Polarity Detection in Presence of Metaphorical Language

The basic idea of this task was the polarity detection (positive or negative or neutral) of tweets containing expressions such as irony, metaphors, sarcasm. The proposed semantic opinion-mining engines had to be assessed according to precision, recall and F-measure computed on the confusion matrix of detected polarity values (positive OR negative) for each tweet of the evaluation dataset. Figure 2 shows an example of the output schema for task1.

Dataset were composed by three thousands of tweets collected from Twitter and already classified with [positive,negative,neutral] polarity values. The manual annotation of each tweet has been performed using Crowdflower[10].

3.3   Task 3: Aspect-Based Sentiment Analysis

Aspect-Based sentiment analysis looks for a binary polarity value associated to aspects extracted from a certain topic. Whereas task 1 and 2 ask for an overall polarity value for a, let's say, given review on a hotel, this task asks for a positive or negative value for aspects of the hotel (rooms' quality, cleanness, food, etc.). Submitted systems are evaluated for the aspect extraction and the performed polarity detection through a precision-recall analysis similarly as performed during SemEval 2016 Task 5[11]. Figure 3 shows an example of the output schema for task 3.

---

[10] https://www.crowdflower.com/.
[11] http://alt.qcri.org/semeval2016/task5/.

```xml
<?xml version="1.0" encoding="UTF-8" standalone="yes"?>
<Sentences>
    <sentence id="apparel_0">
        <text>
        I just love working for 6.5 hours without a break or anything.
        Especially when I'm on my period and have awful cramps.
        </text>
        <polarity>
        negative
        </polarity>
    </sentence>
    <sentence id="apparel_1">
        <text>
        I literally love Stephen A smith haha he's hilarious
        </text>
        <polarity>
        positive
        </polarity>
    </sentence>
</Sentences>
```

Fig. 2. Task 2 output example. Input is the same without the polarity tag.

```xml
<?xml version="1.0" encoding="UTF-8" standalone="yes"?>
<Review rid="1">
    <sentences>
        <sentence id="348:0">
            <text>Most everything is fine with this machine: speed, capacity, build.</text>
            <Opinions>
                <Opinion aspect="MACHINE" polarity="positive"/>
            </Opinions>
        </sentence>
        <sentence id="348:1">
            <text>The only thing I don't understand is that the resolution of the
            screen isn't high enough for some pages, such as Yahoo!Mail.
            </text>
            <Opinions>
                <Opinion aspect="SCREEN" polarity="negative"/>
            </Opinions>
        </sentence>
        <sentence id="277:2">
            <text>The screen takes some getting use to, because it is smaller
            than the laptop.</text>
            <Opinions>
                <Opinion aspect="SCREEN" polarity="negative"/>
            </Opinions>
        </sentence>
    </sentences>
</Review>
```

Fig. 3. Task 3 output example. Input is the same without the opinion tag and its descendant nodes.

The training and test sets were composed by, respectively, 5,058 and 891 sentences coming from three different domains:

- Laptop, (3,048 sentences for training and 728 for testing);
- Restaurant, (2,000 sentences for training);
- Hotel, (163 sentences for testing).

The hotel domain has been chosen to check the efficiency, effectiveness and flexibility of the submitted systems.

3.4   Task 4: Semantic Sentiment Retrieval

Task 4 was related to the retrieval of documents according to certain opinion-based queries. Basically, this task involves Information Retrieval (detect features of given entities), Named Entity Recognition (e.g. detect names of hotels or restaurants), Sentiment Analysis (at sentence level or topic level). Figure 4 shows an example of the input format for task 4.

```
<?xml version="1.0" encoding="UTF-8" standalone="yes"?>
<documents>
    <document id="0">
        <text>So far so good. My wife just loves the new Samsung S5: the display is
        and the colors are very brilliant. However, further memory is necessary for
        everything.</text>
    </document>
    <document id="1">
        <text>All the LG G3 have problems with videos: they often are not able to c
        with tv and when they can, the quality of the image is poor. The only stron
        is the amount of memory coming from the factory.</text>
    </document>
    <document id="2">
        <text>The team behind a project to build one of the world's largest telesco
        Monday it has chosen Spain's Canary Islands in the Atlantic Ocean as a poss
        The decision follows opposition from Native Hawaiians and environmentalists
        constructing the so-called Thirty Meter Telescope (TMT), which would cost $
        volcano on Hawaii's Big Island.</text>
    </document>
</documents>
```

Fig. 4. Task 4 input example.

Figure 5 shows an example of input query. Figure 6 indicates the output related to the input query of Fig. 5.

```
<?xml version="1.0" encoding="UTF-8" standalone="yes"?>
<queries>
    <query id="q0">
        <text>Documents talking about smartphone display.</text>
    </query>
</queries>
```

Fig. 5. Task 4 query format example.

We built from scratch the entire dataset (train + test) and 2 experts validated the annotations computing the relevance of documents according to the Normalized Discounted Cumulated Gain measure. We provided the script to compute Normalized Discounted Cumulated Gain measure within the website of the challenge. The first 20 documents returned by each participant were manually judged.

```xml
<?xml version="1.0" encoding="UTF-8" standalone="yes"?>
<Ranks>
        <query id="q0">
                <position value="1" documentId="0"/>
                <position value="2" documentId="1"/>
        </query>
</Ranks>
```

Fig. 6. Task 4 output format example.

3.5   Task 5: Frame Entities Identification

Frame entities identification task concerns the capability of the challengers' systems to detect the entities involved in a typical opinion sentence depending on their role: holders, topics, opinion concepts. To given an example, let us consider the following sentence: *The mayor is loved by the people in the city, but he has been criticised by the state government* (taken from [10]). The entities that need to be recognised are: *the people* and *state government* as opinion holders, *is loved* and *has been criticized* as opinion concepts, and *The mayor* as a topic of the opinion.

An example of annotations for the sentence above is given in Fig. 7.

```xml
<?xml version="1.0" encoding="UTF-8" standalone="yes"?>
<Sentences>
        <sentence id="348:0">
                <text>The mayor is loved by the people in the city,
                but he has been criticized by the state government.
                </text>
                <Frames>
                        <Frame>
                                <holder start="22" end="32" value="the people"/>
                                <topic start="0" end="9" value="The mayor"/>
                                <opinion start="10" end="18" value="is loved"/>
                                <polarity>positive</polarity>
                        </Frame>
                        <Frame>
                                <holder start="76" end="96" value="the state government"/>
                                <topic start="0" end="9" value="The mayor"/>
                                <opinion start="53" end="72" value="has been criticized"/>
                                <polarity>negative</polarity>
                        </Frame>
                </Frames>
        </sentence>
</Sentences>
```

Fig. 7. Task 5 annotated sentence example.

3.6   Task 6: Subjectivity and Objectivity Detection

This task has commonly been defined as it follows: given a text, classify it into objective or subjective. Basically, an objective sentence does not contain any opinion within it whereas subjective text does. The proposed engines were strongly encouraged to use semantic web solutions, best practices and technologies to solve this task or to indicate how the semantics (even if implicitly adopted) was employed in their methods. An example of the output format of this task is shown in Fig. 8.

```
<?xml version="1.0" encoding="UTF-8" standalone="yes"?>
  <Sentences>
      <sentence id="348:0">
          <text>The mayor is loved by the people in the city.
          </text>
          <value>objective</value>
      </sentence>
      <sentence id="348:0">
          <text>The mayor he has been elected by many voters.
          </text>
          <value>subjective</value>
      </sentence>
  </Sentences>
```

Fig. 8. Task 6 output format example.

Input was the same without the value tag. Training set and test set have be collected from the web, manually annotating 200 documents as subjective or objective. 2–4 experts validated the annotations for this task.

3.7   The Most Innovative Approach Task

The system using in the most innovative way common-sense knowledge and semantics would win the most innovative approach task. We would also take into account the usability of the system, the design of the user interface (if applicable), and multi-language capabilities.

4   Submitted Systems

This year we received 6 expressions of interest but had to discourage one of them because the semantics was not employed at all. The challenge chairs used the Google group mentioned in the introduction section as a forum to explain tasks and requirements needed for the challenge. Details related to the challenge were thus published much ahead of time to let interested researchers know about its status. Thanks to the Google group we have leveraged, there were not any delays during the submission phase. The five submission we received, along their details (title, authors, targeted tasks), are listed in Table 1.

As we had only one participant to task 3 which also applied for task 1 and four more systems that addressed task 1, we considered only task 1 for the competition among the challengers.

Table 1. The systems participating at the fourth edition of the Semantic Sentiment Analysis challenge and the tasks they targeted.

System	Task 1	Task 3	Most Inn. Approach
Mattia Atzeni, Amna Dridi and Diego Reforgiato Recupero **Fine-Grained Sentiment Analysis on Financial Microblogs and News Headlines** [1]	X		X
Marco Federici **A Knowledge-based Approach For Aspect-Based Opinion Mining** [4]	X	X	X
Giulio Petrucci **The IRMUDOSA System at ESWC-2017 Challenge on Semantic Sentiment Analysis** [11]	X		X
Andi Rexha **Exploiting Propositions for Opinion Mining** [18]	X		X
Walid Iguider and Diego Reforgiato Recupero **Language Independent Sentiment Analysis of theShukran Social Network using Apache Spark** [8]	X		X
Giulio Petrucci and Mauro Dragoni **An Information Retrieval-based System For Multi-Domain Sentiment Analysis** [12]	X		X

5   Results

A week before the ESWC conference, the evaluation dataset (the one that contained the sentences only) for Task 1 was published. Participants had to run their systems and send to the challenge chairs their results by the next two days. Computing the accuracy was pretty straightforward as accuracy scripts were already prepared and available to download within the website of the challenge. In the following, we will show the results of the participants' systems.

5.1   Task 1

In Table 2 we show the precision-recall analysis of the four systems competing for Task 1. The system of *Mattia Atzeni, Amna Dridi and Diego Reforgiato Recupero* had the best f-measure and, therefore, was awarded with a Springer voucher of the value of 125 euros, as the winner of the task. To note that two systems have been disqualified because the output format was not compliant with that provided within the challenge task instructions. That is why they are not included in the table.

Table 2. Precision-recall analysis and winners for Task 1.

System	F-Mesure
Mattia Atzeni, Amna Dridi and Diego Reforgiato Recupero **Fine-Grained Sentiment Analysis on Financial Microblogs and News Headlines** [1]	0.8675
Marco Federici **A Knowledge-based Approach For Aspect-Based Opinion Mining** [4]	0.8424
Walid Iguider and Diego Reforgiato Recupero **Language Independent Sentiment Analysis of theShukran Social Network using Apache Spark** [8]	0.8378
Giulio Petrucci **The IRMUDOSA System at ESWC-2017 Challenge on Semantic Sentiment Analysis** [11]	0.8112

5.2   The Most Innovative Approach Task

The Innovation Prize, consisting of a Springer voucher of 125 euros, was awarded to *Marco Federici* with the presented contribution "A Knowledge-based Approach For Aspect-Based Opinion Mining". This system proposed a flexible and innovative way for combining several knowledge bases for addressing the sentiment analysis task. Hence, we decided to assign the award to it.

6   Conclusions

The Semantic Sentiment Analysis challenge at ESWC2017 followed the success of the first three editions and attracted people within the research and industry world from the semantic web community and traditional Sentiment Analysis and natural language processing techniques. In general, researchers coming from the Sentiment Analysis world become curious and familiar with Semantic Web resources and systems and embed them in their existing methods in order to provide higher accuracy.

Our challenge was coupled with a related workshop where participants were suggested to submit a research work explaining the theory behind their method and how the semantics they employed was effectively used. The workshop was a full day event and attracted around 15 people from research and industry.

This year the organizers of the challenge came from an Italian university, an Italian company and a Singaporean university. The three of them are very actively engaged with sentiment analysis research and technology and published and developed several resources, software and papers within that domain. The rationale behind that was to attract researchers and industries from all around the world and have them competing on common tasks of sentiment analysis exploiting Semantic Web technologies. Although only two tasks were targeted, results we obtained of the winning systems were impressive. During the related workshop there was a constructive discussion related to the participants to the challenge and several suggestions were given in order to further improve the precision of those systems, which have also been strongly suggested to participate to other challenge with higher number of participants (e.g. SemEval each year proposes the polarity detection task). Although the number of participants in our challenge were not many, and this is mostly due to the constraints of the Semantic Web technologies that the submitted system had to employ, we aimed at giving advise and suggestions to the few number of participants so that they may compete in other known challenges (e.g. SemEval) with the advantage of exploiting Semantic Web technologies against the systems that still use classical statistical approaches.

We will propose again the dual event challenge-workshop to further provide suggestions and tips to researchers that would like to improve the accuracy of their sentiment analysis methods exploiting Semantic Web technologies and best practices.

Last but not least, we will keep exploiting the Google group we have set up a couple of years ago for dissemination and promotion activities which currently counts 164 members from all around the world.

Acknowledgement. Challenge Organizers want to thank Springer for supporting the provided awards also for this year edition.

Moreover, The authors gratefully acknowledge Sardinia Regional Government for the financial support (Convenzione triennale tra la Fondazione di Sardegna e gli Atenei Sardi Regione Sardegna - L.R. 7/2007 annualit 2016 - DGR 28/21 del 17.05.201, CUP: F72F16003030002).

Moreover, the research leading to these results has received funding from the European Union Horizon 2020 the Framework Programme for Research and Innovation (2014-2020) under grant agreement 643808 Project MARIO Managing active and healthy aging with use of caring service robots.

References

1. Atzeni, M., Dridi, A., Recupero, D.R.: Fine-grained sentiment analysis on financial microblogs and news headlines. In: Solanki and Dragoni [20]
2. Cambria, E., Das, D., Bandyopadhyay, S., Feraco, A.: A Practical Guide to Sentiment Analysis, 1st edn. Springer Publishing Company, Incorporated, Heidelberg (2017). doi:10.1007/978-3-319-55394-8
3. Consoli, S., Gangemi, A., Nuzzolese, A.G., Reforgiato Recupero, D., Spampinato, D.: Extraction of topics-events semantic relationships for opinion propagation in sentiment analysis. In: Proceedings of Extended Semantic Web Conference (ESWC), Crete, GR (2014)
4. Federici, M.: A knowledge-based approach for aspect-based opinion mining. In: Solanki and Dragoni [20]
5. Gandon, F., Cabrio, E., Stankovic, M., Zimmermann, A. (eds.): SemWebEval 2015. CCIS, vol. 548. Springer, Cham (2015). doi:10.1007/978-3-319-25518-7
6. Gangemi, A., Presutti, V., Reforgiato-Recupero, D.: Frame-based detection of opinion holders and topics: a model and a tool. IEEE Comput. Intell. Mag. **9**(1), 20–30 (2014)

7. Gangemi, A., Alani, H., Nissim, M., Cambria, E., Reforgiato Recupero, D., Lanfranchi, V., Kauppinen, T.: Joint Proceedings of the 1st Workshop on Semantic Sentiment Analysis (SSA 2014), and the Workshop on Social Media and Linked Data for Emergency Response (smile 2014) Co-located with 11th European Semantic web Conference (eswc 2014), Crete, Greece, May 25th, 2014 (2014). http:// ceur-ws.org/Vol-1329/

8. Iguider, W., Reforgiato Recupero, D.: Language independent sentiment analysis of the shukran social network using apache spark. In: Solanki and Dragoni [20]

9. Blitzer, J., Dredze, M., Pereira, F.: Biographies, bollywood, boom-boxes and blenders: domain adaptation for sentiment classification. In: Association of Computational Linguistics (ACL) (2007)

10. Liu, B.: Sentiment Analysis and Opinion Mining. Synthesis Lectures on Human Language Technologies. Morgan & Claypool Publishers, San Rafael (2012)

11. Petrucci, G.: The irmudosa system at eswc-2017 challenge on semantic sentiment analysis. In: Solanki and Dragoni [20]

12. Petrucci, G., Dragoni, M.: An information retrieval-based system for multi-domain sentiment analysis. In: Solanki and Dragoni [20]

13. Presutti, V., et al. (eds.): SemWebEval 2014. CCIS, vol. 475. Springer, Cham (2014). doi:10.1007/978-3-319-12024-9

14. Reforgiato Recupero, D., Cambria, E.: ESWC'14 challenge on concept-level sentiment analysis. In: Presutti, V., et al. (eds.) SemWebEval 2014. CCIS, vol. 475, pp. 3–20. Springer, Cham (2014). doi:10.1007/978-3-319-12024-9_1

15. Recupero, D.R., Dragoni, M., Presutti, V.: ESWC 15 challenge on concept-level sentiment analysis. In: Gandon, F., Cabrio, E., Stankovic, M., Zimmermann, A. (eds.) SemWebEval 2015. CCIS, vol. 548, pp. 211–222. Springer, Cham (2015). doi:10.1007/978-3-319-25518-7_18

16. Dragoni, M., Reforgiato Recupero, D.: Challenge on fine-grained sentiment analysis within ESWC2016. In: Sack, H., Dietze, S., Tordai, A., Lange, C. (eds.) SemWebEval 2016. CCIS, vol. 641, pp. 79–94. Springer, Cham (2016). doi:10.1007/ 978-3-319-46565-4_6

17. Recupero, D.R., Presutti, V., Consoli, S., Gangemi, A., Nuzzolese, A.: Sentilo: frame-based sentiment analysis. Cognit. Comput. 7(2), 211–225 (2014)

18. Rexha, A.: Exploiting propositions for opinion mining. In: Solanki and Dragoni [20]

19. Sack, H., Dietze, S., Tordai, A., Lange, C. (eds.): SemWebEval 2016. CCIS, vol. 641. Springer, Cham (2016). doi:10.1007/978-3-319-46565-4

20. Solanki, M., Dragoni, M. (eds.): The Semantic Web: ESWC 2017 Challenges. Communications in Computer and Information Science. Springer, Heidelberg (2017). doi:10.1007/978-3-319-58451-5

Fine-Grained Sentiment Analysis on Financial Microblogs and News Headlines

Mattia Atzeni[✉], Amna Dridi, and Diego Reforgiato Recupero

Università Degli Studi di Cagliari, Department of Mathematics
and Computer Science, Via Ospedale 72, 09124 Cagliari, Italy
ma.atzeni12@studenti.unica.it,
{amna,diego.reforgiato}@unica.it

Abstract. Sentiment analysis in the financial domain is quickly becoming a prominent research topic as it provides a powerful method to predict market dynamics. In this work, we leverage advances in Semantic Web area to develop a fine-grained approach to predict real-valued sentiment scores. We compare several classifiers trained on two different datasets. The first dataset consists of microblog messages focusing on stock market events, while the second one consists of financially relevant news headlines crawled from different sources on the Internet. We test our approach using several feature sets including lexical features, semantic features and a combination of lexical and semantic features. Experimental results show that the proposed approach allows achieving an accuracy level of more than 72%.

Keywords: Sentiment analysis · Financial domain · Stock market prediction · Frame semantics · Microblogs · News · BabelNet · Regression

1 Introduction

In recent years, sentiment analysis has drawn a lot of research interest and has been applied in different domains. The financial domain is particularly relevant, as it has been shown that news and media can deeply affect the market fluctuations [7]. As shown in [8], indeed, positive news usually has a good impact on markets and generally tends to increase optimism. Thus, textual information processing has become a powerful tool to predict market dynamics.

Sentiment analysis in the financial domain has been applied for a wide range of economic and financial fields, such as market prediction [9,13], analyzing consumer's attitudes towards certain brands [6,10] or determining the financial blogger's sentiment towards companies and their stock [12]. Most of the work in sentiment analysis in the financial domain has focused on data collected from finance news. For instance, Ahmad et al. have developed methods for identifying positive and negative news from financial news streams [1].

However, sentiment analysis in the financial domain is still in its early stages, due probably to its interdisciplinary nature. Research in this area, indeed, seldom

© Springer International Publishing AG 2017
M. Dragoni et al. (Eds.): SemWebEval 2017, CCIS 769, pp. 124–128, 2017.
https://doi.org/10.1007/978-3-319-69146-6_11

even makes use of recent advances in the Semantic Web landscape. In this work, we present a system which aims to fill this gap by using *Framester* [5] as a hub between several distinct linguistic resources such as *FrameNet* [2], *WordNet* [4] *BabelNet* [11] and many other well-known resources. The system makes use of lexical features and semantic features extracted by *Framester* to train a classifier on two datasets taken from SemEval 2017 Task 5. The first dataset consists of microblog messages, while the second one consists of sentences taken from news headlines as well as news text. We have tested several classifiers and several combinations of the input features, including approaches based solely on lexical features or semantic features and approaches based on a combination of lexical and semantic features. We finally evaluate the proposed system, which allows to get an accuracy level of more than 72%.

2   System Description

The system has been implemented in Scala, using Big Data technologies such as Apache Spark to parallelize the computation on several processors. For each message, several feature vectors are prepared to test the learning algorithms on different combinations of the input features. The features can be divided in three main categories, namely lexical features, semantic features and a combination of lexical and semantic features.

For lexical features, we used *n-grams*. Each message is first converted to lower case, then the system performs tokenization and lemmatization using Stanford CoreNLP. Next, stop words are removed and we compute *n-grams*. The resulting lexical features consist of *unigrams*, *bigrams* and *3-grams* for each message in the dataset.

Semantic features are extracted using the Framester APIs. Two types of semantic features have been used: *(i) BabelNet synsets*, which are a set of synonyms in different languages grouped by BabelNet and *(ii) Semantic frames*. Both semantic replacement and semantic augmentation methods have been tested.

All the learning algorithms we tested have shown to perform much better when using only a subset of the input features. The features are selected using a metric proposed in [3]. The metric defines the correlation of a word w to the sentiment scores of a set of messages M as follows:

$$c(w, M) = \frac{1}{|M|} \sum_{m \in M} \left\{ I(w, m) \cdot \left(S(m) - \frac{1}{|M|} \sum_{m' \in M} S(m') \right) \right\}$$

where $S(m)$ is the sentiment score associated with message m and $I(w, m)$ is a function that outputs 1 if m contains the word w and outputs -1 otherwise. Intuitively, words which are more likely to appear in messages with a positive sentiment score will have a positive value for the correlation metric, while words that are more likely to appear in messages with a negative value of the sentiment score will receive a negative value for the correlation metric. This metric can be generalized and it is applied separately to unigrams, bigrams, 3-grams, BN synsets and Semantic Frames, to select only the features that are the most significant in determining the polarity of the message.

The last step is to apply TF-IDF scaling to get the final feature vectors that will be used by the classifiers to predict the polarity of the message. The system makes use of Support Vector Machine regression, trained with n-grams, BN synsets and Semantic Frames, as this approach allows achieving the best results, according to our experiments.

3   Data Description

We performed our experiments on two datasets taken from SemEval 2017 Task 5. The first dataset is a collection of 1694 financially relevant microblog messages focusing on stock market events. This messages are either exchanged via the StockTwits microblogging platform or via Twitter. Usually, StockTwits messages contain references to company stock symbols, which are called cashtags. Each message is labeled with a real value denoting the sentiment towards the cashtag. This value is in the range of −1 (very negative/bearish) to +1 (very positive/bullish), with 0 denoting neutral sentiment. Beside the full text of the message, the dataset contains also an attribute called spans, which is a list of strings from the message which are the most significant in expressing sentiment.

The second dataset consists of 1142 sentences taken from headlines as well as news text, crawled from different sources, such as Yahoo Finance. In this case too, each message is labeled with a real value between −1 to +1, denoting the sentiment towards a specific company.

Figure 1 shows the sentiment score distribution for the two datasets.

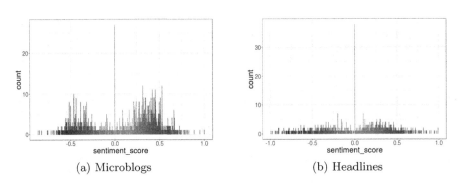

(a) Microblogs   (b) Headlines

Fig. 1. Sentiment score distribution for microblogs and headlines

4   Experimental Results

Five learning algorithms have been tested: Random Forest, Linear Regression, Lasso Regression, Ridge Regression and Support Vector Machine regression. The experimental results have been achieved using 10-fold cross validation and cosine similarity. As the sentiment scores predicted by systems lie on a continuous scale between -1 and 1, cosine enables us to compare the degree of agreement between gold standard and predicted results. At the same time, the scores predicted by the system do not need to be identical to the gold standard in order to achieve a good evaluation result. Table 1 shows the results on microblogs, while Table 2 shows the results on headlines.

Table 1. 10-fold cross validation results on microblogs.

	RF	LinearR	LassoR	RidgeR	SVR
n-grams	**0.680**	0.718	0.582	0.718	0.712
Semantic frames	0.444	0.383	0.314	0.383	0.383
BN synsets	0.570	0.663	0.467	0.662	0.654
BN synsets + semantic frames	0.572	0.660	0.476	0.659	0.661
n-grams + semantic frames	0.674	0.717	0.583	0.717	0.715
n-grams + BN synsets	0.679	**0.725**	0.590	**0.725**	0.724
n-grams + BN synsets + semantic frames	0.675	0.722	**0.592**	0.722	**0.726**

Table 2. 10-fold cross validation results on headlines.

	RF	LinearR	LassoR	RidgeR	SVR
n-grams	**0.563**	**0.633**	**0.516**	**0.634**	0.647
Semantic frames	0.337	0.329	0.294	0.392	0.328
BN synsets	0.456	0.550	0.384	0.551	0.530
BN synsets + semantic frames	0.465	0.540	0.389	0.541	0.543
n-grams + semantic frames	0.559	0.624	0.506	0.624	0.625
n-grams + BN synsets	0.556	0.626	0.510	0.627	0.649
n-grams + BN synsets + semantic frames	0.554	0.619	0.502	0.620	**0.655**

5   Conclusions

Sentiment analysis in the financial domain is an interesting task, as several studies have shown that it can be useful for predicting future values of stock prices. This work introduces an approach that brings together natural language processing and Semantic Web, to predict a real-valued sentiment score. We have considered three main categories of features: lexical features, semantic features and a combination of the lexical and semantic features. Five learning algorithms have been compared. The system succeeded in achieving the accuracy level of more than 72% when the training model was boosted by semantics.

References

1. Ahmad, K., Cheng, D., Almas, Y.: Multi-lingual sentiment analysis of financial news streams. In: Proceedings of the 1st International Conference on Grid in Finance. Palermo, Italy (2006)

2. Baker, C.F., Fillmore, C.J., Lowe, J.B.: The berkeley framenet project. In: Proceedings of the 17th International Conference on Computational Linguistics COLING 1998, Association for Computational Linguistics, vol. 1, pp. 86–90. Stroudsburg (1998). http://dx.doi.org/10.3115/980451.980860

3. Drake, A., Ringger, E.K., Ventura, D.: Sentiment regression: using real-valued scores to summarize overall document sentiment. In: ICSC, pp. 152–157. IEEE Computer Society (2008)

4. Fellbaum, C. (ed.): WordNet: An Electronic Lexical Database. MIT Press, Cambridge (1998)

5. Gangemi, A., Alam, M., Asprino, L., Presutti, V., Recupero, D.R.: Framester: a wide coverage linguistic linked data hub. In: Blomqvist, E., Ciancarini, P., Poggi, F., Vitali, F. (eds.) EKAW 2016. LNCS (LNAI), vol. 10024, pp. 239–254. Springer, Cham (2016). doi:10.1007/978-3-319-49004-5_16

6. Ghiassi, M., Skinner, J., Zimbra, D.: Twitter brand sentiment analysis: a hybrid system using n-gram analysis and dynamic artificial neural network. Expert Syst. Appl. **40**(16), 6266–6282 (2013). http://dx.doi.org/10.1016/j.eswa.2013.05.057

7. Goonatilake, R., Herath, S.: The volatility of the stock market and news. Int. Res. J. Finan. Econ. **3**(11), 53–65 (2007)

8. Van de Kauter, M., Breesch, D., Hoste, V.: Fine-grained analysis of explicit and implicit sentiment in financial news articles. Expert Syst. Appl. **42**(11), 4999–5010 (2015). http://dx.doi.org/10.1016/j.eswa.2015.02.007

9. Khadjeh Nassirtoussi, A., Aghabozorgi, S., Ying Wah, T., Ngo, D.C.L.: Review: text mining for market prediction: a systematic review. Expert Syst. Appl. **41**(16), 7653–7670 (2014). http://dx.doi.org/10.1016/j.eswa.2014.06.009

10. Mostafa, M.M.: More than words: social networks' text mining for consumer brand sentiments. Expert Syst. Appl. **40**(10), 4241–4251 (2013). http://dx.doi.org/10.1016/j.eswa.2013.01.019

11. Navigli, R., Ponzetto, S.P.: Babelnet: the automatic construction, evaluation and application of a wide-coverage multilingual semantic network. Artif. Intell. **193**, 217–250 (2012). http://dx.doi.org/10.1016/j.artint.2012.07.001

12. O'Hare, N., Davy, M., Bermingham, A., Ferguson, P., Sheridan, P., Gurrin, C., Smeaton, A.F.: Topic-dependent sentiment analysis of financial blogs. In: Proceedings of the 1st International CIKM Workshop on Topic-sentiment Analysis for Mass Opinion, pp. 9–16. TSA 2009. ACM, New York (2009). http://doi.acm.org/10.1145/1651461.1651464

13. Sprenger, T.O., Tumasjan, A., Sandner, P.G., Welpe, I.M.: Tweets and trades: the information content of stock microblogs. Eur. Financ. Manag. **20**(5), 926–957 (2014). http://dx.doi.org/10.1111/j.1468-036X.2013.12007.x

Language Independent Sentiment Analysis
of the Shukran Social Network
Using Apache Spark

Walid Iguider[(✉)] and Diego Reforgiato Recupero

Department of Mathematics and Computer Science,
University of Cagliari, Cagliari, Italy
walid.iguider@gmail.com, diego.reforgiato@unica.it

Abstract. This paper describes theShukran Sentiment Analysis system.
TheShukran is a social network micro-blogging service that allows users
posting photos or videos and descriptions of their daily life activities.
This social network rapidly gained a large amount of users. It provides
people from different cultures and countries the possibility to share in
different languages their stories, ideas, opinions, and news from their real
life, and makes the cultural diversity the center of relationships between
its users. Sentiment analysis aims to extract the opinion of the public
about some topic by processing text data. One of its several tasks, the
polarity detection, aims at categorizing the elements in a dataset (sentences, posts, etc.) into classes such as positive, negative and neutral.
In the system we propose, and that represents the sentiment analysis
core engine of theShukran social network, we will detect the original
language of users posts, translate them into English and evaluate their
sentiment (whether positive, negative or neutral). We propose the use of
a Naive Bayes classifier and SentiWordNet and SenticNet for the sentiment evaluation. The language detection and translation are performed
using TextBlob, a Python library for processing textual data.

Keywords: Sentiment analysis · Natural language processing · Polarity
detection · Naive Bayes · SentiWordNet · SenticNet

1 Introduction

Recently, knowing authors' opinion polarity on a specific topic in natural language
texts published in social networks and mini-blogs is becoming the area of interest
of either companies and research, as the topics posted could contain political or
religious issues, product, movie or music review and a lot of other content that can
contain relevant information about the authors opinion [1–3,9].

The social network we are considering, $theShukran^1$, is a place where people
express freely their daily life, in their mother language, without any influence

---

[1] http://www.theshukran.com/.

© Springer International Publishing AG 2017
M. Dragoni et al. (Eds.): SemWebEval 2017, CCIS 769, pp. 129–132, 2017.
https://doi.org/10.1007/978-3-319-69146-6_12

of a third party which makes the content reflecting objectively the opinions of users.

The user generated content on the social networks can contain a variety of important market research information and opinions, through which we can predict new economic opportunities and risks at an early stage. [3]

Nevertheless, the content is generally unstructured and posted in different languages. The most used languages in *theShukran*, for example, are Urdu and Arabic since in a first place, theShukran was dedicated to the Muslim community. This makes it harder the processing as it requires to collect training datasets or lexical resources for each covered language. As we are interested in the polarity detection problem, we target Task 1 of the ESWC-17 Challenge on Semantic Sentiment Analysis.

Our idea is to use English dataset to train the classifier using existing Semantic Web resources and to translate every comment on the social network into English; then we proceed with the classification.

We have built a Naive Bayesian classifier, described as a function that can give the correlation between a certain feature and a class.

The dataset we used in our system is Twitter Sentiment Analysis Dataset[2]. As it is built from social network data, it makes a good corpus to learn from in our case.

2   The Classification Work-Flow

We have built a Naive Bayes classifier trained on our dataset. It contains 1,578,627 classified tweets. Each row is marked as 1 for positive sentiment and 0 for negative sentiment. From the whole dataset 788,442 sentence are marked as negative whereas 790,185 are marked as positive.

2.1   Preprocessing

Due to the fact that we will have to deal with social network data, obviously, in different languages, the process of text cleaning has to be preceded by automatic language recognition and the translation of all the documents into English.

Automatic Language Recognition and Translation: We have chosen TextBlob[3], a Python library for processing textual data. It is free and it does not have any usage limitations. Furthermore, it provides a simple API for diving into common natural language processing (NLP) tasks such as part-of-speech tagging, noun phrase extraction, classification, translation and other.

---

[2] http://thinknook.com/wp-content/uploads/2012/09/Sentiment-Analysis-Dataset. zip.

[3] https://textblob.readthedocs.io/en/dev/.

Tokenization: Tokenization is the process to split sentences into words and forming bag of words. [4]

In this phase the steps we performed were: (i) Breaking the sentence into "tokens"; (ii) Lowercasing the tokens; (iii) Removing punctuation; (iv) Removing stop-words; (v) Stemming the tokens.

Finally, as translation tools are never perfect and resulting sentence might contain some grammar error, we acted at word level instead of sentence level and removed the words that were not included in SentiWordNet[4] and compound expressions not included in SenticNet[5].

SentiWordNet: SentiWordNet is a public lexical resource, containing opinion information on terms extracted from the WordNet database. It is widely used in a variety of applications, such as emotional (or sentiment) analysis, and opinion mining. It also provides sentiment scores of positive, negative scores for each word, by using semi-supervised learning, automatic annotation of WordNet, and a random-walk approach [6,7].

SenticNet: SenticNet [8] leverages semantics and linguistics to perform sentiment analysis tasks. It is (i) a concept-level knowledge base, (ii) a multi-disciplinary framework, (iii) a private company. SenticNet provides semantic information, sentics and polarity associated with more than 50,000 natural language concepts. Semantics are concepts that are most semantically-related to the input concept, sentics are emotion categorization values expressed in terms of four affective dimensions (Pleasantness, Attention, Sensitivity, and Aptitude) and polarity is floating number between -1 and $+1$ (where -1 is extreme negativity and $+1$ is extreme positivity).

2.2 Feature Extraction

In this phase, the tokenized text is transformed into a feature vector that feeds the Naive Bayes classifier. We used a term frequency feature extractor that maps each raw feature into an index (term) by applying a hash function. Then it calculates term frequencies based on the mapped indexes.

3 The Application

The application is developed using Python, Apache Spark and MLlib [5]. **Apache Spark** is a popular open-source platform for large-scale data processing that is well-suited for iterative machine learning tasks over big data. **MLlib** is Spark's open-source distributed machine learning library. It provides efficient functionality for a wide range of learning settings and includes several underlying statistical, optimization, and linear algebra primitives. Shipped with Spark,

---

[4] http://sentiwordnet.isti.cnr.it/.

[5] http://sentic.net/.

MLlib supports several languages and provides a high-level API that leverages Spark's rich ecosystem to simplify the development of end-to-end machine learning pipelines.

The application we developed is able to take in argument a sentence in any language and translates it into English, then it evaluates its sentiment.

4   Conclusion and Future Work

We have presented the sentiment analysis system core engine of *theShukran* social network that tackles the problem of polarity detection. It employs Naive Bayes classification, Apache Spark for big data processing, MLib for machine learning with Spark and uses SentiWordNet and SenticNet to remove words or compound expressions not expressing emotions or opinions. The system is a participant to the task 1 of the ESWC-17 Challenge on Semantic Sentiment Analysis. As theShukran social network contains images and videos with related descriptions and comments, we are working on extending our system to provide live polarity detection as soon as a user writes something.

References

1. Mewari, R., Singh, A., Srivastava, A.: Article: opinion mining techniques on social media data. Int. J. Comput. Appl. **118**(6), 39–44 (2015)
2. Pang, B., Lee, L.: Opinion mining and sentiment analysis. Found. Trends Inf. Retrieval **2**(1–2), 1–135 (2008)
3. Petz, G., Karpowicz, M., Fürschuß, H., Auinger, A., Stříteský, V., Holzinger, A.: Opinion mining on the web 2.0 – characteristics of user generated content and their impacts. In: Holzinger, A., Pasi, G. (eds.) HCI-KDD 2013. LNCS, vol. 7947, pp. 35–46. Springer, Heidelberg (2013). doi:10.1007/978-3-642-39146-0_4
4. ChandraKala, S., Sindhu, C.: Opinion mining and sentiment classification: a survey. ICTACT J. Soft Comput. **3**(1), 420–427 (2012)
5. Meng, X., Bradley, J., Yavuz, B., Sparks, E., Venkataraman, S., Liu, D., Freeman, J., Tsai, D., Amde, M., Owen, S., Xin, D., Xin, R., Franklin, M.J., Zadeh, R., Zaharia, M., Talwalkar, A.: MLlib: Machine learning in Apache Spark. J. Mach. Learn. Res. **17**, 1–7 (2016)
6. Cambria, E., Hussain, A.: Sentic Computing: Techniques, Tools and Applications, vol. 2. Springer, Netherlands (2012). doi:10.1007/978-94-007-5070-8
7. Baccianella, S., Esuli, A., Sebastiani, F.: SentiWordNet 3.0: an enhanced lexical resource for sentiment analysis and opinion mining. In: LREC (2010)
8. Cambria, E., Poria, S., Bajpai, R., Schuller, B.: SenticNet 4: a semantic resource for sentiment analysis based on conceptual primitives. In: Proceedings of COLING 2016, The 26th International Conference on Computational Linguistics: Technical Papers, pp. 2666–2677 (2016)
9. Recupero, D.R., Presutti, V., Consoli, S., Gangemi, A., Nuzzolese, A.G.: Sentilo: frame-based sentiment analysis. Cognitive Comput. **7**(2), 211–225 (2015)

Aspect-Based Opinion Mining
Using Knowledge Bases

Marco Federici[1] and Mauro Dragoni[2(✉)]

[1] Universitá di Trento, Trento, Italy
federici@fbk.eu
[2] Fondazione Bruno Kessler, Trento, Italy
dragoni@fbk.eu

Abstract. In the last decade, the focus of the Opinion Mining field moved to detection of the pairs "aspect-polarity" instead of limiting approaches in the computation of the general polarity of a text. In this work, we propose an aspect-based opinion mining system based on the use of semantic resources for the extraction of the aspects from a text and for the computation of their polarities. The proposed system participated at the third edition of the Semantic Sentiment Analysis (SSA) challenge took place during ESWC 2017 achieving the runner-up place in the Task #2 concerning the aspect-based sentiment analysis. Moreover, a further evaluation performed on the SemEval 2015 benchmarks demonstrated the feasibility of the proposed approach.

1 Introduction and Related Work

Opinion Mining is a natural language processing (NLP) task that aims to classify documents according to their opinion (polarity) on a given subject [1]. This task has created a considerable interest due to its wide applications in different domains: marketing, politics, social sciences, etc. Generally, the polarity of a document is computed by analyzing the expressions contained in the full text without distinguishing which are the subjects of each opinion. In the last decade, the research in the opinion mining field focused on the "aspect-based opinion mining" [2] consisting in the extraction of all subjects ("aspects") from documents and the opinions that are associated with them.

For clarification, let us consider the following example:

Yesterday, I bought a new smartphone.
The quality of the display is very good, but the buttery lasts too little.

In the sentence above, we may identify three aspects: "smartphone", "display", and "battery". As the reader may see, each aspect has a different opinion associated with it. The list below summarizes such associations:

- "display" → "very good"
- "battery" → "too little"

M. Dragoni et al. (Eds.): SemWebEval 2017, CCIS 769, pp. 133–147, 2017.
https://doi.org/10.1007/978-3-319-69146-6_13

- "smartphone" → no explicit opinions, therefore polarity can be inferred by averaging the opinions associated with all other aspects.

The topic of aspect-based sentiment analysis has been explored under different perspectives. A comprehensive review of the last available systems can be found in the proceedings of SemEval 2015[1].

The paper is structured as follows. Section 2 briefly provides an overview of the aspect extraction task. Section 3 introduces the background knowledge used during the development of the system. Section 4 presents the underlying NLP layer upon which it has been developed the system described Sect. 5. Sections 6 and 7 shows the results obtained on the ESWC 2016 SSA challenge and on the SemEval 2015 benchmark, respectively. Finally, Sect. 8 provide a description about how the tasks of the challenge have been addressed and it concludes the paper.

2   Related Work

The topic of sentiment analysis has been studied extensively in the literature [3], where several techniques have been proposed and validated.

Machine learning techniques are the most common approaches used for addressing this problem, given that any existing supervised methods can be applied to sentiment classification. For instance, in [4], the authors compared the performance of Naive-Bayes, Maximum Entropy, and Support Vector Machines in sentiment analysis on different features like considering only unigrams, bigrams, combination of both, incorporating parts of speech and position information or by taking only adjectives. Moreover, beside the use of standard machine learning method, researchers have also proposed several custom techniques specifically for sentiment classification, like the use of adapted score function based on the evaluation of positive or negative words in product reviews [5], as well as by defining weighting schemata for enhancing classification accuracy [6].

An obstacle to research in this direction is the need of labeled training data, whose preparation is a time-consuming activity. Therefore, in order to reduce the labeling effort, opinion words have been used for training procedures. In [7,8], the authors used opinion words to label portions of informative examples for training the classifiers. Opinion words have been exploited also for improving the accuracy of sentiment classification, as presented in [9], where a framework incorporating lexical knowledge in supervised learning to enhance accuracy has been proposed. Opinion words have been used also for unsupervised learning approaches like the one presented in [10].

Another research direction concerns the exploitation of discourse-analysis techniques. [11] discusses some discourse-based supervised and unsupervised approaches for opinion analysis; while in [12], the authors present an approach to identify discourse relations.

---

[1] http://alt.qcri.org/semeval2015/cdrom/pdf/SemEval082.pdf.

The approaches presented above are applied at the document-level [13–16], i.e., the polarity value is assigned to the entire document content. However, in some case, for improving the accuracy of the sentiment classification, a more fine-grained analysis of a document is needed. Hence, the sentiment classification of the single sentences, has to be performed. In the literature, we may find approaches ranging from the use of fuzzy logic [17–19] to the use of aggregation techniques [20] for computing the score aggregation of opinion words. In the case of sentence-level sentiment classification, two different sub-tasks have to be addressed: (i) to determine if the sentence is subjective or objective, and (ii) in the case that the sentence is subjective, to determine if the opinion expressed in the sentence is positive, negative, or neutral. The task of classifying a sentence as subjective or objective, called "subjectivity classification", has been widely discussed in the literature [21–24] and systems implementing the capabilities of identifying opinion's holder, target, and polarity have been presented [25]. Once subjective sentences are identified, the same methods as for sentiment classification may be applied. For example, in [26] the authors consider gradable adjectives for sentiment spotting; while in [27–29] the authors built models to identify some specific types of opinions.

In the last years, with the growth of product reviews, the use of sentiment analysis techniques was the perfect floor for validating them in marketing activities [30]. However, the issue of improving the ability of detecting the different opinions concerning the same product expressed in the same review became a challenging problem. Such a task has been faced by introducing "aspect" extraction approaches that were able to extract, from each sentence, which is the aspect the opinion refers to. In the literature, many approaches have been proposed: conditional random fields (CRF) [31], hidden Markov models (HMM) [32], sequential rule mining [33], dependency tree kernels [34], clustering [35], and genetic algorithms [36]. In [37,38], two methods were proposed to extract both opinion words and aspects simultaneously by exploiting some syntactic relations of opinion words and aspects.

A particular attention should be given also to the application of sentiment analysis in social networks [39,40]. More and more often, people use social networks for expressing their moods concerning their last purchase or, in general, about new products. Such a social network environment opened up new challenges due to the different ways people express their opinions, as described by [41,42], who mention "noisy data" as one of the biggest hurdles in analyzing social network texts.

One of the first studies on sentiment analysis on micro-blogging websites has been discussed in [43], where the authors present a distant supervision-based approach for sentiment classification.

At the same time, the social dimension of the Web opens up the opportunity to combine computer science and social sciences to better recognize, interpret, and process opinions and sentiments expressed over it. Such multi-disciplinary approach has been called *sentic computing* [44]. Application domains where sentic computing has already shown its potential are the cognitive-inspired classification of images [45], of texts in natural language, and of handwritten text [46].

Finally, an interesting recent research direction is domain adaptation, as it has been shown that sentiment classification is highly sensitive to the domain from which the training data is extracted. A classifier trained using opinionated documents from one domain often performs poorly when it is applied or tested on opinionated documents from another domain, as we demonstrated through the example presented in Sect. 1. The reason is that words and even language constructs used in different domains for expressing opinions can be quite different. To make matters worse, the same word in one domain may have positive connotations, but in another domain may have negative ones; therefore, domain adaptation is needed. In the literature, different approaches related to the Multi-Domain sentiment analysis have been proposed. Briefly, two main categories may be identified: (i) the transfer of learned classifiers across different domains [47–49], and (ii) the use of propagation of labels through graph structures [17,50–52].

All approaches presented above are based on the use of statistical techniques for building sentiment models. The exploitation of semantic information is not taken into account. In this work, we proposed a first version of a semantic-based approach preserving the semantic relationships between the terms of each sentence in order to exploit them either for building the model and for estimating document polarity. The proposed approach, falling into the multi-domain sentiment analysis category, instead of using pre-determined polarity information associated with terms, it learns them directly from domain-specific documents. Such documents are used for training the models used by the system.

3   Preliminaries

The system is implemented on top of a background knowledge used for representing the linguistic connections between "concepts" described in several resources. Below, it is possible to find the list of such resources and the links where further information about them may be found.

WordNet[2] [53] is one of the most important resource available to researchers in the field of text analysis, computational linguistics, and many related areas. In the implemented system, WordNet has been used as starting point for the construction of the semantic graph used by the system (see Sect. 5) However, due to some coverage limitations occurring in WordNet, it has been extended by linking further terms coming from the Roget's Thesaurus [54].

SenticNet[3] [55] is a publicly available resource for opinion mining that exploits both Artificial Intelligence and Semantic Web techniques to infer the polarity associated with common-sense concepts and represent it in a semantic-aware format. In particular, SenticNet uses dimensionality reduction to calculate the affective valence of a set of Open Mind concepts and represent it in a machine-accessible and machine-processable format.

---

[2] https://wordnet.princeton.edu/.

[3] http://sentic.net/.

General Inquirer dictionary[4] [56] is an English-language dictionary containing almost 12,000 elements associated with their polarity in different contexts. Such dictionary is the result of the integration between the "Harvard" and the "Lasswell" general-purpose dictionaries as well as a dictionary of categories define by the dictionary creators. When necessary, for ambiguous words, specific polarity for each sense is specified.

4   The Underlying NLP Layer

The presented system has been implemented on top of existing Natural Language Processing libraries. In particular, it uses different functionalities offered by the Stanford NLP Library.

WordNet (see Footnote 2) [53] resource is used together with Stanford's part of speech annotation to detect compound nouns. Lists of consecutive nouns and word sequences contained in Wordnet compound nouns vocabulary are merged into a single word in order to force Stanford library to consider them as a single unit during the following phases. The entire text is then fed to the co-reference resolution module to compute pronoun references which are stored in an index-reference map. Details about the textual analysis are provided in Sect. 5.

The next operation consists in detecting which word expresses polarity within each sentence. To achieve this task *SenticNet*, *General Inquirer dictionary* and *MPQA* sentiment lexicons have been used.

While SenticNet expresses polarity values in the continuous range from -1 to 1, the other two resources been normalized: the General Inquirer words have positive values of polarity if they belong to the "Positiv" class while negative if they belong to "Negativ" one, zero otherwise, similarly, MPQA "polarity" labels are used to infer a numerical values. Only words with a non-zero polarity value in at least one resource are considered as opinion words (e.g. word "third" is not present in MPQA and SenticNet and has a 0 value according to General Inquirer, consequently, it is not a valid opinion word; on the other hand, word "huge" has a positive 0.069 value according to SenticNet, a negative value in MPQA and 0 value according to General Inquirer, therefore, it is a possible opinion word even if lexicons express contrasting values). Every noun (single or complex) is considered an aspect as long as it's connected to at least one opinion and it's not in the stopword list. This list has been created starting from the "Onix" text retrieval engine stopwords list[5] and it contains words without a specific meaning (such as "thing") and special characters.

Opinions associated with pronouns are connected to the aspect they are referring to; instead, if pronouns reference can't be resolved, they are both discarded.

---

[4] http://www.wjh.harvard.edu/~inquirer/spreadsheet_guide.htm.
[5] The used stopwords list is available at http://www.lextek.com/manuals/onix/stopwords1.html.

The main task of the system is, then, represented by connecting opinions with possible aspects. Two different approaches have been tested with a few variants. The first one relies on the syntactic tree while the second one is based on grammar dependencies.

The sentence "I enjoyed the screen resolution, it's amazing for such a cheap laptop." has been used to underline differences in connection techniques.

The preliminary phase merges words "screen" and "resolution" into a single word "Screenresolution" because they are consecutive nouns. Co-reference resolution module extracts a relation between "it" and "Screenresolution". This relation is stored so that every possible opinion that would be connected to "it" will be connected to "Screenresolution" instead. Figure 1 shows the syntax tree while Fig. 2 represents the grammar relation graph generated starting from the example sentence. Both structures have been computed using Stanford NLP modules ("parse", "depparse").

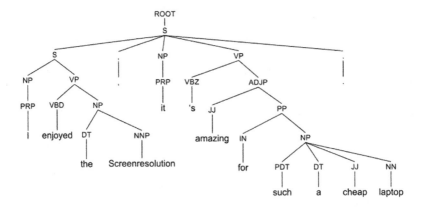

Fig. 1. Example of syntax tree.

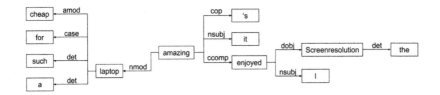

Fig. 2. Example of the grammar relations graph.

5   The Implemented System

The aspect extraction component is based on a six-phases approach as presented below.

Phase 1. Sentences are given as input to the Stanford NLP Library[6] and they are annotated with part of speech (POS) tags in order to detect nouns, adjectives, and pronouns.

Phase 2. Tokens annotated as adjectives are considered for computing opinion scores, while sequences of one or more consecutive nouns (for example "support" tagged as "NN" followed by "team" tagged "NN" as well) and complex linguistic structures recognized through the use of Wordnet (for example "hard" annotated as "JJ" and "disk" annotated as "NN") are aggregated and marked as potential aspects. This step is shown in Fig. 3

Phase 3. Co-reference resolution is applied for resolving pronouns co-references between nouns. Example about how co-reference is applied is shown in Fig. 3.

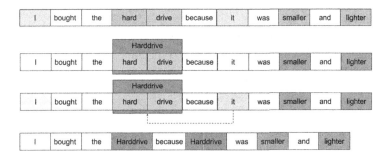

Fig. 3. Example of noun aggregation and co-reference resolution.

Phase 4. After the aggregation of compound names, we changed the sentences by replacing compound names with single tokens to ensure that they are considered as single entities during the opinion resolution phase. This way, it will be possible to exchange each pronoun with the corresponding label of the aspect they are referring to.

Phase 5. Stanford Parser is used for generating a syntax tree that is exploited in the last phase for associating opinions with aspects. Concerning the definition of the associations between aspects and opinions, during preliminary testing activities, we tried different approaches. Among them:

- each aspect has been connected with each opinion contained in the same sentence, where as sentence delimiters, we used the markers "S", "SBAR", and "FRAG" detected in the parsed tree;
- if an opinion is expressed in a sentence without nouns, such an opinion has been associated with the aspects belonging to the same noun phrase only.

---

[6] http://stanfordnlp.github.io/CoreNLP/index.html.

Example of the generated parsed tree is shown in Fig. 4.

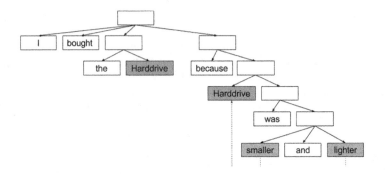

Fig. 4. Example of generated parse tree.

Phase 6. Finally, aspects without associated opinions are discarded, while remaining ones are stored. Example is shown in Fig. 5.

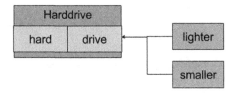

Fig. 5. Example of opinion association.

6   ESWC-2016 SSA Challenge Task #2: Aspect-Based Sentiment Analysis

The expected output of this task was a set of aspects of the reviewed product and a binary polarity value associated to each of such aspects. So, for example, while for classic binary polarity inference task an overall polarity (positive or negative) was expected for a review about a mobile phone, this task required a set of aspects (such as *speaker, touchscreen, camera*, etc.) and a polarity value (positive or negative) associated with each of such aspects. Engines were assessed according to both aspect extraction and aspect polarity detection using precision, recall, f-measure, and accuracy similarly as performed during the first edition of the Concept-Level Sentiment Analysis challenge held during ESWC2014 and re-proposed at SemEval 2015 Task 12[7]. Please refer to SemEval 2016 Task 5[8] for details on the precision-recall analysis. Figure 6 shows an example of the output schema for Task #2.

---

[7] http://www.alt.qcri.org/semeval2015/task12/.
[8] http://alt.qcri.org/semeval2016/task5/.

```
<?xml version="1.0" encoding="UTF-8" standalone="yes"?>
<Review rid="1">
    <sentences>
        <sentence id="348:0">
            <text>Most everything is fine with this machine: speed, capacity, build.</text>
                <Opinions>
                    <Opinion aspect="MACHINE" polarity="positive"/>
                </Opinions>
        </sentence>
        <sentence id="348:1">
            <text>The only thing I don't understand is that the resolution of the
                screen isn't high enough for some pages, such as Yahoo!Mail.
            </text>
            <Opinions>
                <Opinion aspect="SCREEN" polarity="negative"/>
            </Opinions>
        </sentence>
        <sentence id="277:2">
            <text>The screen takes some getting use to, because it is smaller
                than the laptop.</text>
            <Opinions>
                <Opinion aspect="SCREEN" polarity="negative"/>
            </Opinions>
        </sentence>
    </sentences>
</Review>
```

Fig. 6. Task #2 output example. Input is the same without the opinion tag and its descendant nodes.

The training set was composed by 5,058 sentences coming from two different domains: "Laptop" (3,048 sentences) and "Restaurant" (2,000 sentences). While, the test set was composed by 891 sentences coming from the "Laptop" (728 sentences) and "Hotels" (163 sentence). The reason for which we decided to use the "Hotels" domain in the test set with respect to the "Restaurant" one was to observe the capability of the participant systems to be general purpose with respect to the training set.

7   System Evaluation and Error Analysis

The system has been tested on two aspect-based sentiment analysis datasets by following the "Semi-Open" setting of the DRANZIERA protocol [57]:

D1 The SemEval 2015 Task 12 training set benchmark, consisting in sentences belonging to the "Laptop" and "Restaurant" domains.
D2 The ESWC2016 Benchmark on Semantic Sentiment Analysis test set, consisting in sentences belonging to the "Laptop" and "Hotels" domains.

To compute results, a notion of correctness has to be introduced: if the extracted aspects is equal, contained or contains the correct one, it's considered to be correct (for example if the extracted aspect is "screen", while the annotated one is "screen of the computer" or vice versa, the result of the system is considered to correct).

Tables 1 shows the results obtained on each dataset; while, Table 2 shows the full results of Task #2 participants.

Table 1. Results obtained by the presented system on the SemEval 2015 Task 12 dataset and on the test set adopted for the challenge.

Dataset	Precision	Recall	F-measure	Polarity accuracy
D1	0.39969	0.39478	0.39722	0.91720
D2	0.34820	0.35745	0.35276	0.84925

Table 2. Precision-recall analysis and winners for Task 2.

System	Precision	Recall	F-measure	Accuracy
Soufian Jebbara and Philipp Cimiano **Aspect-based sentiment analysis using a two-step neural network architectures**	0.41471	0.45196	0.43253	0.87356
Marco Federici and Mauro Dragoni **A knowledge-based approach for aspect-based opinion mining**	0.34820	0.35745	0.35276	0.84925
Andi Rexha, Mark Kröll Mauro Dragoni and Roman Kern **Exploiting propositions for opinion mining**	N/A	N/A	N/A	N/A

Figure 7 shows an analysis of error cases. Values have been computed according to the first 100 sentences of the "Laptop" dataset.

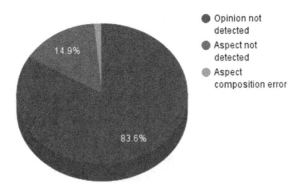

Fig. 7. Overall error analysis.

The majority of false negatives are given by the impossibility to detect opinions expressed by verbs. For example, in the sentence "I generally like this place" or more complex expressions "tech support would not fix the problem unless I bought your plan for $150 plus".

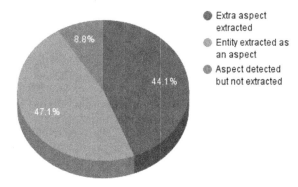

Fig. 8. Specific error analysis on aspect extraction.

Other issues are correlated to the association algorithm. Figure 8 shows the specific error analysis related to the extraction of aspects, always computed on the same 100 sentences of the "Laptop" dataset.

Even if the syntax-tree-based approach tends to produce a significant number of true positives, relationships are often imprecise. A relevant example is represented by the sentence "I was extremely happy with the OS itself." in the "Laptop" dataset. The approach connects the opinion adjective "happy" with the potential aspect "OS", correctly recognized as an aspect in the sentence.

A relevant part of false positives are generated due to the incapability of discriminating aspects from the entity itself. In facts, almost half of them consists in associations between opinion words and the entity reviewed that are correct. However, they must not be considered during the aspect extraction task (for example the aspect "laptop" in the example sentence should not be considered according to the definition of aspect).

8 Conclusions

In this paper, we presented the system submitted to the third edition of the Semantic Sentiment Analysis run during ESWC 2016. The system participated only at Task #2 and obtained the second place. The problem of detecting aspects in sentences is very relevant in the sentiment analysis community. Further work in this direction will be performed by starting from the analysis of the errors provided in the previous section.

References

1. Pang, B., Lee, L., Vaithyanathan, S.: Thumbs up? Sentiment classification using machine learning techniques. In: Proceedings of EMNLP, Philadelphia, pp. 79–86. Association for Computational Linguistics, July 2002

2. Hu, M., Liu, B.: Mining and summarizing customer reviews. In: Proceedings of the Tenth ACM SIGKDD International Conference on Knowledge Discovery and Data Mining, pp. 168–177. ACM (2004)
3. Liu, B., Zhang, L.: A survey of opinion mining and sentiment analysis. In: Aggarwal, C.C., Zhai, C.X. (eds.) Mining Text Data, pp. 415–463. Springer, Boston (2012). doi:10.1007/978-1-4614-3223-4_13
4. Pang, B., Lee, L.: A sentimental education: sentiment analysis using subjectivity summarization based on minimum cuts. In: ACL, pp. 271–278 (2004)
5. Dave, K., Lawrence, S., Pennock, D.M.: Mining the peanut gallery: opinion extraction and semantic classification of product reviews. In: WWW, pp. 519–528 (2003)
6. Paltoglou, G., Thelwall, M.: A study of information retrieval weighting schemes for sentiment analysis. In: ACL, pp. 1386–1395 (2010)
7. Tan, S., Wang, Y., Cheng, X.: Combining learn-based and lexicon-based techniques for sentiment detection without using labeled examples. In: SIGIR, pp. 743–744 (2008)
8. Qiu, L., Zhang, W., Hu, C., Zhao, K.: SELC: a self-supervised model for sentiment classification. In: CIKM, pp. 929–936 (2009)
9. Melville, P., Gryc, W., Lawrence, R.D.: Sentiment analysis of blogs by combining lexical knowledge with text classification. In: KDD, pp. 1275–1284 (2009)
10. Taboada, M., Brooke, J., Tofiloski, M., Voll, K.D., Stede, M.: Lexicon-based methods for sentiment analysis. Comput. Linguist. **37**(2), 267–307 (2011)
11. Somasundaran, S.: Discourse-level relations for opinion analysis. Ph.D. thesis, University of Pittsburgh (2010)
12. Wang, H., Zhou, G.: Topic-driven multi-document summarization. In: IALP, pp. 195–198 (2010)
13. Dragoni, M.: SHELLFBK: an information retrieval-based system for multi-domain sentiment analysis. In: Proceedings of the 9th International Workshop on Semantic Evaluation (SemEval 2015), Denver, pp. 502–509. Association for Computational Linguistics, June 2015
14. Petrucci, G., Dragoni, M.: An information retrieval-based system for multi-domain sentiment analysis. In: Gandon, F., Cabrio, E., Stankovic, M., Zimmermann, A. (eds.) SemWebEval 2015. CCIS, vol. 548, pp. 234–243. Springer, Cham (2015). doi:10.1007/978-3-319-25518-7_20
15. Rexha, A., Kröll, M., Dragoni, M., Kern, R.: Exploiting propositions for opinion mining. In: Sack, H., Dietze, S., Tordai, A., Lange, C. (eds.) SemWebEval 2016. CCIS, vol. 641, pp. 121–125. Springer, Cham (2016). doi:10.1007/978-3-319-46565-4_9
16. Federici, M., Dragoni, M.: A knowledge-based approach for aspect-based opinion mining. In: Sack, H., Dietze, S., Tordai, A., Lange, C. (eds.) SemWebEval 2016. CCIS, vol. 641, pp. 141–152. Springer, Cham (2016). doi:10.1007/978-3-319-46565-4_11
17. Dragoni, M., Tettamanzi, A.G., da Costa Pereira, C.: Propagating and aggregating fuzzy polarities for concept-level sentiment analysis. Cogn. Comput. **7**(2), 186–197 (2015)
18. Dragoni, M., Tettamanzi, A.G.B., da Costa Pereira, C.: A fuzzy system for concept-level sentiment analysis. In: Presutti, V., et al. (eds.) SemWebEval 2014. CCIS, vol. 475, pp. 21–27. Springer, Cham (2014). doi:10.1007/978-3-319-12024-9_2
19. Petrucci, G., Dragoni, M.: The IRMUDOSA system at ESWC-2016 challenge on semantic sentiment analysis. In: Sack, H., Dietze, S., Tordai, A., Lange, C. (eds.) SemWebEval 2016. CCIS, vol. 641, pp. 126–140. Springer, Cham (2016). doi:10.1007/978-3-319-46565-4_10

20. da Costa Pereira, C., Dragoni, M., Pasi, G.: A prioritized "And" aggregation operator for multidimensional relevance assessment. In: Serra, R., Cucchiara, R. (eds.) AI*IA 2009. LNCS, vol. 5883, pp. 72–81. Springer, Heidelberg (2009). doi:10.1007/978-3-642-10291-2_8

21. Federici, M., Dragoni, M.: Towards unsupervised approaches for aspects extraction. In: Dragoni, M., Recupero, D.R., Denecke, K., Deng, Y., Declerck, T. (eds.) Joint Proceedings of the 2th Workshop on Emotions, Modality, Sentiment Analysis and the Semantic Web and the 1st International Workshop on Extraction and Processing of Rich Semantics from Medical Texts Co-located with ESWC 2016, Heraklion, 29 May 2016. CEUR Workshop Proceedings, vol. 1613. CEUR-WS.org (2016)

22. Federici, M., Dragoni, M.: A branching strategy for unsupervised aspect-based sentiment analysis. In: Dragoni, M., Recupero, D.R. (eds.) Proceedings of the 3rd International Workshop on Emotions, Modality, Sentiment Analysis and the Semantic Web Co-located with 14th ESWC 2017, Portroz, 28 May 2017. CEUR Workshop Proceedings, vol. 1874. CEUR-WS.org (2017)

23. Riloff, E., Patwardhan, S., Wiebe, J.: Feature subsumption for opinion analysis. In: EMNLP, pp. 440–448 (2006)

24. Wilson, T., Wiebe, J., Hwa, R.: Recognizing strong and weak opinion clauses. Comput. Intell. 22(2), 73–99 (2006)

25. Palmero Aprosio, A., Corcoglioniti, F., Dragoni, M., Rospocher, M.: Supervised opinion frames detection with RAID. In: Gandon, F., Cabrio, E., Stankovic, M., Zimmermann, A. (eds.) SemWebEval 2015. CCIS, vol. 548, pp. 251–263. Springer, Cham (2015). doi:10.1007/978-3-319-25518-7_22

26. Hatzivassiloglou, V., Wiebe, J.: Effects of adjective orientation and gradability on sentence subjectivity. In: COLING, pp. 299–305 (2000)

27. Kim, S.M., Hovy, E.H.: Crystal: analyzing predictive opinions on the web. In: EMNLP-CoNLL, pp. 1056–1064 (2007)

28. Rexha, A., Kröll, M., Dragoni, M., Kern, R.: Polarity classification for target phrases in tweets: a Word2Vec approach. In: Sack, H., Rizzo, G., Steinmetz, N., Mladenić, D., Auer, S., Lange, C. (eds.) ESWC 2016. LNCS, vol. 9989, pp. 217–223. Springer, Cham (2016). doi:10.1007/978-3-319-47602-5_40

29. Rexha, A., Kröll, M., Kern, R., Dragoni, M.: An embedding approach for microblog polarity classification. In: Dragoni, M., Recupero, D.R. (eds.) Proceedings of the 3rd International Workshop on Emotions, Modality, Sentiment Analysis and the Semantic Web Co-located with 14th ESWC 2017, Portroz, 28 May 2017. CEUR Workshop Proceedings, vol. 1874. CEUR-WS.org (2017)

30. Dragoni, M., Reforgiato Recupero, D.: Challenge on fine-grained sentiment analysis within ESWC2016. In: Sack, H., Dietze, S., Tordai, A., Lange, C. (eds.) SemWebEval 2016. CCIS, vol. 641, pp. 79–94. Springer, Cham (2016). doi:10.1007/978-3-319-46565-4_6

31. Jakob, N., Gurevych, I.: Extracting opinion targets in a single and cross-domain setting with conditional random fields. In: EMNLP, pp. 1035–1045 (2010)

32. Jin, W., Ho, H.H., Srihari, R.K.: Opinionminer: a novel machine learning system for web opinion mining and extraction. In: KDD, pp. 1195–1204 (2009)

33. Liu, B., Hu, M., Cheng, J.: Opinion observer: analyzing and comparing opinions on the web. In: WWW, pp. 342–351 (2005)

34. Wu, Y., Zhang, Q., Huang, X., Wu, L.: Phrase dependency parsing for opinion mining. In: EMNLP, pp. 1533–1541 (2009)

35. Su, Q., Xu, X., Guo, H., Guo, Z., Wu, X., Zhang, X., Swen, B., Su, Z.: Hidden sentiment association in Chinese web opinion mining. In: WWW, pp. 959–968 (2008)
36. Dragoni, M., Azzini, A., Tettamanzi, A.G.B.: A novel similarity-based crossover for artificial neural network evolution. In: Schaefer, R., Cotta, C., Kołodziej, J., Rudolph, G. (eds.) PPSN 2010. LNCS, vol. 6238, pp. 344–353. Springer, Heidelberg (2010). doi:10.1007/978-3-642-15844-5_35
37. Qiu, G., Liu, B., Bu, J., Chen, C.: Opinion word expansion and target extraction through double propagation. Comput. Linguist. **37**(1), 9–27 (2011)
38. Dragoni, M.: Extracting linguistic features from opinion data streams for multi-domain sentiment analysis. In: Dragoni, M., Recupero, D.R. (eds.) Proceedings of the 3rd International Workshop on Emotions, Modality, Sentiment Analysis and the Semantic Web Co-located with 14th ESWC 2017, Portroz, 28 May 2017. CEUR Workshop Proceedings, vol. 1874. CEUR-WS.org (2017)
39. Dragoni, M.: A three-phase approach for exploiting opinion mining in computational advertising. IEEE Intell. Syst. **32**(3), 21–27 (2017)
40. Dragoni, M., Petrucci, G.: A neural word embeddings approach for multi-domain sentiment analysis. IEEE Trans. Affect. Comput. **PP**(99), 1 (2017)
41. Barbosa, L., Feng, J.: Robust sentiment detection on Twitter from biased and noisy data. In: COLING (Posters), pp. 36–44 (2010)
42. Bermingham, A., Smeaton, A.F.: Classifying sentiment in microblogs: is brevity an advantage? In: CIKM, pp. 1833–1836 (2010)
43. Go, A., Bhayani, R., Huang, L.: Twitter sentiment classification using distant supervision. CS224N Project Report, Standford University (2009)
44. Cambria, E., Hussain, A.: Sentic Computing: A Common-Sense-Based Framework for Concept-Level Sentiment Analysis. Springer, Cham (2015). doi:10.1007/978-3-319-23654-4
45. Cambria, E., Hussain, A.: Sentic album: content-, concept-, and context-based online personal photo management system. Cogn. Comput. **4**(4), 477–496 (2012)
46. Wang, Q.F., Cambria, E., Liu, C.L., Hussain, A.: Common sense knowledge for handwritten Chinese recognition. Cogn. Comput. **5**(2), 234–242 (2013)
47. Blitzer, J., Dredze, M., Pereira, F.: Biographies, bollywood, boom-boxes and blenders: domain adaptation for sentiment classification. In: ACL, pp. 187–205 (2007)
48. Pan, S.J., Ni, X., Sun, J.T., Yang, Q., Chen, Z.: Cross-domain sentiment classification via spectral feature alignment. In: WWW, pp. 751–760 (2010)
49. Yoshida, Y., Hirao, T., Iwata, T., Nagata, M., Matsumoto, Y.: Transfer learning for multiple-domain sentiment analysis—identifying domain dependent/independent word polarity. In: AAAI, pp. 1286–1291 (2011)
50. Ponomareva, N., Thelwall, M.: Semi-supervised vs. cross-domain graphs for sentiment analysis. In: RANLP, pp. 571–578 (2013)
51. Huang, S., Niu, Z., Shi, C.: Automatic construction of domain-specific sentiment lexicon based on constrained label propagation. Knowl. Based Syst. **56**, 191–200 (2014)
52. Dragoni, M., da Costa Pereira, C., Tettamanzi, A.G.B., Villata, S.: Smack: an argumentation framework for opinion mining. In: Kambhampati, S. (ed.) Proceedings of the Twenty-Fifth International Joint Conference on Artificial Intelligence (IJCAI 2016), New York, 9–15 July 2016, pp. 4242–4243. IJCAI/AAAI Press (2016)
53. Fellbaum, C.: WordNet: An Electronic Lexical Database. MIT Press, Cambridge (1998)

54. Kipfer, B.A.: Roget's 21st century thesaurus, 3rd edn. May 2005
55. Cambria, E., Speer, R., Havasi, C., Hussain, A.: SenticNet: a publicly available semantic resource for opinion mining. In: AAAI Fall Symposium: Commonsense Knowledge (2010)
56. Stone, P.J., Dunphy, D., Smith, M.: The General Inquirer: A Computer Approach to Content Analysis. M.I.T Press, Oxford (1966)
57. Dragoni, M., Tettamanzi, A., da Costa Pereira, C.: Dranziera: an evaluation protocol for multi-domain opinion mining. In: Calzolari, N., Choukri, K., Declerck, T., Goggi, S., Grobelnik, M., Maegaard, B., Mariani, J., Mazo, H., Moreno, A., Odijk, J., Piperidis, S. (eds.) Proceedings of the Tenth International Conference on Language Resources and Evaluation (LREC 2016), Paris. European Language Resources Association (ELRA), May 2016

The IRMUDOSA System at ESWC-2017 Challenge on Semantic Sentiment Analysis

Giulio Petrucci[1] and Mauro Dragoni[2(✉)]

[1] Universitá di Trento, Trento, Italy
petrucci@fbk.eu
[2] Fondazione Bruno Kessler, Trento, Italy
dragoni@fbk.eu

Abstract. Multi-Domain opinion mining consists in estimating the polarity of a document by exploiting domain-specific information. One of the main issue of the approaches discussed in literature is their poor capability of being applied on domains that have not been used for building the opinion model. In this paper, we present an approach exploiting the linguistic overlap between domains for building models enabling the estimation of polarities for documents belonging to any other domain. The system implementing such an approach has been presented at the third edition of the Semantic Sentiment Analysis Challenge co-located with ESWC 2017. Fuzzy representation of features polarity supports the modeling of information uncertainty learned from training set and integrated with knowledge extracted from two well-known resources used in the opinion mining field, namely Sentic.Net and the General Inquirer. The proposed technique has been validated on a multi-domain dataset and the results demonstrated the effectiveness of the proposed approach by setting a plausible starting point for future work.

1 Introduction

Opinion mining is a natural language processing task aiming to classify documents according to the opinion (polarity) they express on a given subject [1]. Generally speaking, opinion mining aims at determining the attitude of a speaker or a writer with respect to a topic or the overall tonality of a document. This task has created a considerable interest due to its wide applications. In recent years, the exponential increase of the Web for exchanging public opinions about events, facts, products, etc., has led to an extensive usage of opinion mining approaches, especially for marketing purposes.

Most of the work available in the literature address the opinion mining problem without distinguishing the domains which documents, used for building models, come from. The necessity of investigating this problem from a multi-domain perspective is led by the different influence that a term might have in different contexts. Let us consider the following examples. In the first example, we have an "emotion-based" context where the adjective "cold" is used differently based on the feeling, or mood, of the opinion holder:

M. Dragoni et al. (Eds.): SemWebEval 2017, CCIS 769, pp. 148–165, 2017.
https://doi.org/10.1007/978-3-319-69146-6_14

1. That person always behaves in a very **cold** way with her colleagues.
2. A **cold** drink is the best thing we can drink when the temperature is very hot.

while in the second one, we have a "subjective-based" context where the adjective "small" is used differently based on the product category reviewed by a user:

1. The sideboard is **small** and it is not able to contain a lot of stuff.
2. The **small** dimensions of this decoder allow to move it easily.

In the first context, we considered two different "emotional" situations: in the first one a person is commenting about the behavior of his colleague by using the adjective "cold" with a "negative" polarity. Instead, in the second one, a person is referring to the adjective "cold" in a "positive" way as a good solution for a situation.

Instead, in the second context, we considered the interpretation of texts referring to two different domain: "Furnishings" and "Electronics". In the first one, the polarity of the adjective "small" is, for sure, negative because it highlights an issue of the described item. On the other hand, in the second domain, the polarity of such an adjective may be considered positive.

The multiple facets with which textual information can be analyzed in the context of opinion mining led to the design of approaches creating models able to address this scenario. The idea of adapting terms polarity to different domains emerged only recently [2]. In general, multi-domain opining mining approaches discussed in the literature (surveyed in Sect. 2) focus on building models for transferring information between pairs of domains [3]. While on one hand such approaches allow to propagate specific domain information to other, their drawback is the necessity of building new transfer models any time a new domain has to be addressed. This way, approaches include a poor generalization capability of analyzing text, because transfer models are limited to the N domains used for building the models.

This paper describes our approach exploiting the linguistic overlap between domains for building models enabling the estimation of polarities for documents. Due to this peculiarity, the proposed approach is innovative, to the best of our knowledge, with respect to the state of the art of multi-domain opinion mining.

The rest of the article is structured as follows. Section 2 presents a survey on works about opinion mining either in the single o multi domain environment. Section 3 provides the references to the knowledge resources used in the implementation of the proposed approach described in detail in Sect. 4. Section 5 reports the system evaluation and, finally, Sect. 6 concludes the article.

2   Related Work

The topic of sentiment analysis has been studied extensively in the literature [4], where several techniques have been proposed and validated.

Machine learning techniques are the most common approaches used for addressing this problem, given that any existing supervised methods can be applied to sentiment classification. For instance, in [5], the authors compared the performance of Naive-Bayes, Maximum Entropy, and Support Vector Machines in sentiment analysis on different features like considering only unigrams, bigrams, combination of both, incorporating parts of speech and position information or by taking only adjectives. Moreover, beside the use of standard machine learning method, researchers have also proposed several custom techniques specifically for sentiment classification, like the use of adapted score function based on the evaluation of positive or negative words in product reviews [6], as well as by defining weighting schemata for enhancing classification accuracy [7].

An obstacle to research in this direction is the need of labeled training data, whose preparation is a time-consuming activity. Therefore, in order to reduce the labeling effort, opinion words have been used for training procedures. In [8,9], the authors used opinion words to label portions of informative examples for training the classifiers. Opinion words have been exploited also for improving the accuracy of sentiment classification, as presented in [10], where a framework incorporating lexical knowledge in supervised learning to enhance accuracy has been proposed. Opinion words have been used also for unsupervised learning approaches like the one presented in [11].

Another research direction concerns the exploitation of discourse-analysis techniques. [12] discusses some discourse-based supervised and unsupervised approaches for opinion analysis; while in [13], the authors present an approach to identify discourse relations.

The approaches presented above are applied at the document-level [14–17], i.e., the polarity value is assigned to the entire document content. However, in some case, for improving the accuracy of the sentiment classification, a more fine-grained analysis of a document is needed. Hence, the sentiment classification of the single sentences, has to be performed. In the literature, we may find approaches ranging from the use of fuzzy logic [18–20] to the use of aggregation techniques [21] for computing the score aggregation of opinion words. In the case of sentence-level sentiment classification, two different sub-tasks have to be addressed: (i) to determine if the sentence is subjective or objective, and (ii) in the case that the sentence is subjective, to determine if the opinion expressed in the sentence is positive, negative, or neutral. The task of classifying a sentence as subjective or objective, called "subjectivity classification", has been widely discussed in the literature [22–25] and systems implementing the capabilities of identifying opinion's holder, target, and polarity have been presented [26]. Once subjective sentences are identified, the same methods as for sentiment classification may be applied. For example, in [27] the authors consider gradable adjectives for sentiment spotting; while in [28–30] the authors built models to identify some specific types of opinions.

In the last years, with the growth of product reviews, the use of sentiment analysis techniques was the perfect floor for validating them in marketing activities [31]. However, the issue of improving the ability of detecting the different opinions concerning the same product expressed in the same review became a challenging problem. Such a task has been faced by introducing "aspect" extraction approaches that were able to extract, from each sentence, which is the aspect the opinion refers to. In the literature, many approaches have been proposed: conditional random fields (CRF) [32], hidden Markov models (HMM) [33], sequential rule mining [34], dependency tree kernels [35], clustering [36], and genetic algorithms [37]. In [38,39], two methods were proposed to extract both opinion words and aspects simultaneously by exploiting some syntactic relations of opinion words and aspects.

A particular attention should be given also to the application of sentiment analysis in social networks [39,40]. More and more often, people use social networks for expressing their moods concerning their last purchase or, in general, about new products. Such a social network environment opened up new challenges due to the different ways people express their opinions, as described by [41,42], who mention "noisy data" as one of the biggest hurdles in analyzing social network texts.

One of the first studies on sentiment analysis on micro-blogging websites has been discussed in [43], where the authors present a distant supervision-based approach for sentiment classification.

At the same time, the social dimension of the Web opens up the opportunity to combine computer science and social sciences to better recognize, interpret, and process opinions and sentiments expressed over it. Such multi-disciplinary approach has been called *sentic computing* [44]. Application domains where sentic computing has already shown its potential are the cognitive-inspired classification of images [45], of texts in natural language, and of handwritten text [46].

Finally, an interesting recent research direction is domain adaptation, as it has been shown that sentiment classification is highly sensitive to the domain from which the training data is extracted. A classifier trained using opinionated documents from one domain often performs poorly when it is applied or tested on opinionated documents from another domain, as we demonstrated through the example presented in Sect. 1. The reason is that words and even language constructs used in different domains for expressing opinions can be quite different. To make matters worse, the same word in one domain may have positive connotations, but in another domain may have negative ones; therefore, domain adaptation is needed. In the literature, different approaches related to the Multi-Domain sentiment analysis have been proposed. Briefly, two main categories may be identified: (i) the transfer of learned classifiers across different domains [2,3,47], and (ii) the use of propagation of labels through graph structures [18,48–50].

All approaches presented above are based on the use of statistical techniques for building sentiment models. The exploitation of semantic information is not taken into account. In this work, we proposed a first version of a semantic-based approach preserving the semantic relationships between the terms of each sentence in order to exploit them either for building the model and for estimating document polarity. The proposed approach, falling into the multi-domain sentiment analysis category, instead of using pre-determined polarity information associated with terms, it learns them directly from domain-specific documents. Such documents are used for training the models used by the system.

3   Knowledge Resources

The proposed approach exploits the use of background knowledge for supporting the creation of the multi-domain model used for computing text polarities. Such a background knowledge is composed by two linguistic resources freely available to the research community. Below, we briefly describe them, while in Sect. 4, we present how they have been used for supporting the implementation of the proposed approach.

SenticNet. *SenticNet*[1] [51] is a publicly available resource for opinion mining that exploits both artificial intelligence and semantic Web techniques to infer the polarities associated with common-sense concepts and to represent them in a semantic-aware format. In particular, SenticNet uses dimensionality reduction to calculate the affective valence of a set of Open Mind[2] concepts and it represents them in a machine accessible and processable format. SenticNet contains more than 5,700 polarity concepts (nearly 40% of the Open Mind corpus) and it may be connected with any kind of opinion mining application. For example, after the de-construction of the text into concepts through a semantic parser, SenticNet can be used to associate polarity values to these and, hence, infer the overall polarity of a clause, sentence, paragraph, or document by averaging such values.

General Inquirer. *General Inquirer dictionary*[3] [52] is an English-language dictionary containing almost 12,000 elements associated with their polarity in different contexts. Such dictionary is the result of the integration between the "Harvard" and the "Lasswell" general-purpose dictionaries as well as a dictionary of categories define by the dictionary creators. When necessary, for ambiguous words, specific polarity for each sense is specified. For every words, a set of tags is provided in the dictionary. Among them, only a subset is relevant to the opinion mining topic and have been exploited in this work: "valence categories", "semantic dimensions", "words of pleasure", and "words reflecting presence or lack of emotional expressiveness". Other categories indicating ascriptive social categories rather than references to places have been considered out of the scope of the opinion mining topic and have not been considered in the implementation of the approach.

---

[1] http://sentic.net/.

[2] http://commons.media.mit.edu/en/.

[3] http://www.wjh.harvard.edu/~inquirer/spreadsheet_guide.htm.

4   Method

The main goal of the presented approach is to exploit domain overlap for compensating the lack of knowledge caused by building models using only a snapshot of the reality. When a domain-based model is built, part of the knowledge belonging to such a domain is not included in the model due to its missing in the adopted training set. Besides, when a system has to classify a text belonging to a domain that has not been used for building the model, it is possible that such a model does not contain enough information for estimating text opinion. For this reason, it is necessary to compensate this lack of knowledge by partially exploiting information coming from other domains.

In this section, we describe the steps adopted for building the models and the strategy we implemented for computing the polarity of a text by exploiting domain overlapping.

The model construction process is composed by three steps: (i) features are extracted from each text contained in the training set (Sect. 4.1); then, (ii) a preliminary fuzzy membership functions [53], modeled by using a triangular shape, is computed by analyzing only the explicit information contained in the dataset (Sect. 4.2); and, (iii) this shape is transformed into a trapezoid after a refinement operation performed by compensating the uncertainty inherited by the adoption of a training set, with information coming from external resources: the SenticNet knowledge base and the General Inquirer dictionary (Sect. 4.3).

Finally, when the inference of a text polarity is requested, the usage of the model for estimating text polarities is performed by aggregating the polarities associated with each feature detected from the text to polarize and by computing the final judgment (Sect. 4.4).

4.1   Feature Extraction

During the Feature Extraction (FE) phase, documents are analyzed and significant elements are extracted and used as features for building the model. As feature, we mean every text chunk that may have a meaning in the context of opinion mining and/or in domain detection. The first step that we performed for extracting the features is to parse the content of each document by using the Stanford NLP Parser [54]. The parser has been used for annotating terms with part of speech (POS) tags and for extracting the dependencies tree of each sentence.

Let's consider the following text marked with a positive polarity: "This smartphone is great. The display is awesome and the touch system works very well."

By parser the text, we obtained the analysis of the two detected sentences. Their POS-tagged versions are represented in the lines 1 and 2; while, the dependencies of the first sentence are shown in the lines from 3 to 6 and, finally, the dependencies of the second sentence are shown in the lines from 7 to 17.

```
1.  This/DT smartphone/NN is/VBZ great/JJ ./.
2.  The/DT display/NN is/VBZ awesome/JJ and/CC the/DT
    touch/NN system/NN works/VBZ very/RB well/RB ./.

3.  root ( ROOT-0 , great-4 )
4.  det ( smartphone-2 , This-1 )
5.  nsubj ( great-4 , smartphone-2 )
6.  cop ( great-4 , is-3 )

7.  root ( ROOT-0 , awesome-4 )
8.  det ( display-2 , The-1 )
9.  nsubj ( awesome-4 , display-2 )
10. cop ( awesome-4 , is-3 )
11. cc ( awesome-4 , and-5 )
12. det ( system-8 , the-6 )
13. compound ( system-8 , touch-7 )
14. nsubj ( works-9 , system-8 )
15. conj:and ( awesome-4 , works-9 )
16. advmod ( well-11 , very-10 )
17. advmod ( works-9 , well-11 )
```

From the parser output, we distinguished two type of features for building our model:

- Single concepts feature: nouns, adjectives, and verbs are stored in the model as single features. Nouns are used for building the domain detection component of our model; while, adjectives and verbs are used for building the opinion mining component. By considering as example the dependencies extracted from the first sentence, the term "smartphone" is inserted in the model with the role of supporting the domain detection, while the term "great" is inserted in the model with the role of supporting the definition about how positive polarity is modeled.
- Terms dependency feature: a selection of the dependencies extracted by the parser is stored within the model with the aim of incorporating domain specific contextual knowledge describing (i) how concepts are connected in a particular domain and (ii) how such connections are related to a particular polarity. The kind of dependencies took into account are "noun-adjective", "adjective-verb", "noun-verb", and "adjective-adverb". In the example above, lines containing significant terms dependency features are 5, 9, 10, 14, 16, and 17. From each "term dependency" feature we actually extract further features that are inserted in the model as well. Let's consider as example the dependency at line 16. From the dependency, we extract "well-very", "very-well", "well", and "very".

4.2 Preliminary Learning Phase

The Preliminary Learning (PL) phase aims at estimating the starting polarity and the domain belonging degree (DBD) of each feature. The estimation of these values is done by analyzing only the explicit information provided by the training set.

Concerning the estimation of the feature polarity, this phase allows to define the preliminary fuzzy membership functions representing the polarity of each feature extracted from the training set with respect to the domain containing such polarity. The feature polarity is estimated as:

$$p_i^E(F) = \frac{k_F^i}{T_F^i} \in [-1,1] \qquad \forall i = 1, \ldots, n, \tag{1}$$

where F is the feature taken into account, index i refers to domain D_i which the feature belongs to, n is the number of domains available in the training set, k_F^i is the arithmetic sum of the polarities observed for feature F in the training set restricted to domain D_i, T_F^i is the number of instances of the training set, restricted to domain D_i, in which feature F occurs, and E stays for "estimated". The shape of the fuzzy membership function generated during this phase is a triangle with the top vertex in the coordinates $(x, 1)$, where $x = p_i^{(E)}(F)$ and with the two bottom vertexes in the coordinates $(-1, 0)$ and $(1, 0)$ respectively. The rationale is that while we have one point (x) in which we have full confidence, our uncertainty covers the entire space because we do not have any information concerning the remaining polarity values. At this stage, the types of feature took into account are the terms dependency features and, as single concept features, adjectives and verbs.

Figure 1 shows a picture of the generated fuzzy triangle.

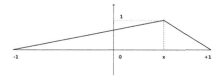

Fig. 1. The fuzzy triangle generated after the preliminary learning phase.

After the polarity estimation, we computed the DBD of each feature. Such a value is exploited during the Polarity Aggregation and Decision Phase (described in Sect. 4.4) for computing the final polarity of a document.

The computation of the DBD is inspired by the well-known TF-IDF model [55] used in information retrieval, where the importance of a term is given by either the frequency of a term in a document contained in an index and the inverse of the number of documents in which such a term occurs.

In our case, the DBD of a feature is computed by summing two factors: the feature frequency associated with the domain in which the feature occurs and the "uniqueness" of the feature with respect to all domains. The domain-frequency is computed as:

$$\text{freq}_i(F) = \frac{k_i(F)}{N_i} \tag{2}$$

where F is the feature taken into account, i is the domain that is analyzed, k_i is the number of times that the feature F occurs in the domain i, and N_i is the total number of features contained in the domain i.

While, the feature uniqueness is computed as:

$$\text{uniq}_i(F) = \frac{\text{freq}_i(F)}{\sum\limits_{i=0}^{n} \text{freq}_i(F)} \tag{3}$$

where F is the feature taken into account and n is the number of domains.

Finally the DBD of each feature is given by:

$$\text{DBD}_i(F) = \text{freq}_i(F) + \text{uniq}_i(F) \tag{4}$$

4.3 Information Refinement Phase

Polarities estimated during the PL phase are refined by exploiting, for each feature, polarities extracted from the resources described in Sect. 3. The rationale behind this choice is to balance the polarity estimated from the training set (that represents only a snapshot of the world) with polarity information that are contained in supervised knowledge bases. When we estimate the polarity value of each feature, two scenarios may happen:

1. the estimated polarity "agrees" (i.e. it has the same orientation) with the one extracted from the knowledge bases;
2. the estimated polarity "disagrees" with the one extracted from the knowledge bases.

In the first case, the estimated polarity confirms, in terms of opinion orientation, what it is represented in the knowledge bases. The representation of this kind of uncertainty will be a tight shape.

On the contrary, in the second case, the estimated polarity is the opposite of what has been extracted from the knowledge bases. In this case, the uncertainty associated with the feature will produce a larger shape. Such a shape will model the contrast between what has been estimated from the training set and what has been defined by experts in the construction of the knowledge bases.

For this reason an Information Refinement (IR) phase is necessary in order to convert this uncertainty in a numerical representation that can be managed by the system.

Assume to have the following values associated to the feature F belonging to the domain i:

- p_s^F, represents the polarity of the feature F extracted from SenticNet;
- p_g^F, represents the polarity of the feature F extracted from the General Inquirer;
- avg_p is the average polarity computed among $p_i^E(F)$, p_s^F, and p_g^F;
- var_p is the variance computed between avg_p and the three polarities values $p_i^E(F)$, p_s^F, and p_g^F.

For "terms dependency" features, that are composed by two terms T_1 and T_2, the values p_s^F and p_g^F are the average of the single polarities computed on T_1 and T_2, respectively.

By starting from these values, the final shape of the inferred fuzzy membership functions, at the end of the IR phase, is a trapezoid whose core consists of the interval between the polarity value learned during the PL phase, $p_i^E(F)$, and avg_p. While, the support of the fuzzy shape is given, on both sides, by the variance var_p.

To sum up, for each domain D_i, $\mu_{F,i}$ is a trapezoid with parameters (a, b, c, d), where

$$a = \min\{p_i^E(F), avg_p\},$$
$$b = \max\{p_i^E(F), avg_p\},$$
$$c = \max\{-1, a - var_p\},$$
$$d = \min\{1, b + var_p\}.$$

The idea here is that the most likely values for the polarity of F for domain D_i are those comprised between the estimated value and average between the estimation of the training algorithm and the polarity values retrieved from supervised knowledge resources. The uncertainty modeled by the fuzzy shape is proportional to the level of "agreement" between the estimated polarity value, and the polarities retrieved from the supervised knowledge bases.

Figure 2 shows a picture of the generated fuzzy trapezoid.

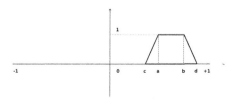

Fig. 2. The fuzzy trapezoid generated after the information refinement phase.

4.4   Polarity Aggregation and Decision Phase

The fuzzy polarities of different features, resulting from the IR phase, are finally aggregated by a fuzzy averaging operator obtained by applying the extension principle [56] in order to compute fuzzy polarities for complex entities, like texts, which consist of a number of features and thus derive, so to speak, their polarity from them. When a crisp polarity value is needed, it may be computed from a fuzzy polarity by applying a defuzzification [57] method.

The operation of computing the entire polarity of a text is done not only on the model describing the domain which the document belongs, but also on the other domains. Indeed, one of the assumption of the proposed approach is to exploit possible linguistic overlaps for compensating missing knowledge of the training set. Therefore, given $p_i(D)$ as the polarity computed on document D with respect to the model of the domain i, and $DBD_i(D)$ as the domain belonging degree of document D to the domain i, we compute the following two vectors:

$$\langle \text{polarity}_D \rangle = [p_0(D), p_1(D), \ldots, p_n(D)] \tag{5}$$

and

$$\langle \text{domain}_D \rangle = [DBD_0(D), DBD_1(D), \ldots, DBD_n(D)] \tag{6}$$

where n is the number of domains contained in the model. The final polarity of a text is then computed by multiplying the two vectors as follow:

$$p_T = \frac{\langle \text{polarity}_D \rangle \times \langle \text{domain}_D \rangle}{N} \tag{7}$$

where N is the number of domains used for building the model.

5 System Evaluation

Here, we present the evaluation procedure adopted for validating the proposed approach.

5.1 The Dataset

The system has been trained by using the DRANZIERA dataset [58]. The dataset is composed by one million reviews crawled from product pages on the Amazon web site[4]. Such reviews belong to twenty different domains, we called in-model domains (IMD): Amazon Instant Video, Automotive, Baby, Beauty, Books, Clothing Accessories, Electronics, Health, Home Kitchen, Movies TV, Music, Office Products, Patio, Pet Supplies, Shoes, Software, Sports Outdoors, Tools Home Improvement, Toys Games, and Video Games.

For each domain, we extracted twenty-five thousands positive and twenty-five thousands negative reviews that have been split in five folds containing five thousand positive and five thousand negative reviews each. This way, the dataset is balanced with respect to either the polarities of the reviews and to the domain which they belong to. The choice between positive and negative documents has been inspired by the strategy used in [2] where reviews with 4 or 5 stars have been marked as positive, while the ones having 1 or 2 stars have been marked as negative.

Besides the twenty domains mentioned above, we used further 7 test sets for measuring the effectiveness of the approach in estimating polarities of texts belonging to domains different from the ones used to build the model, we called out-model domains (OMD). Such domains are: Cell Phones Accessories, Gourmet Foods, Industrial Scientific, Jewelry, Kindle Store, Musical Instruments, and Watches.

---

[4] All the material used for the evaluation and the built models are available at http://goo.gl/pj0nWS.

5.2 Evaluation Procedure

The approach has been evaluated through a 5-cross-fold evaluation procedure. For each execution we measured the precision and the recall and, at the end, we report their averages together with the standard deviation measured over the five executions.

The approach has been compared with three baselines:

- Most Frequent Polarity (MFP): results obtained by guessing always the same polarity for all instances contained in the test set.
- Domain Belonging Polarity (DBP): results obtained by computing the text polarity by using only the information of the domain the text belongs to. This means that the linguistic overlap between domains has not been considered.
- Domain Detection Polarity (DDP): results obtained by computing the text polarity by using only the information of the domain guessed as the most appropriate one for the text to evaluate. This means that the similarity between text content and domain is preferred with respect to the domain used for tagging the text.

The same baselines have been used for evaluating the OMD test sets. In this case, the DBP baseline has not been applied due to the mismatch between the domains used for building the model and the ones contained in the test sets. Each OMD test set has been applied to all five models built, and the scores averaged.

5.3 In-Vitro Results

Here, we show the results of the evaluation campaign conducted for validating the presented approach. Tables 1 and 2 present a summary of the performance obtained by our system and by the three baselines on IMD and OMD, respectively. First column contains the name of the approach, second, third and fourth ones contain the average precision, recall and F1 score computed over all domains; while, the fifth column contains the average standard deviation computed on the F1 score during the cross-fold validation. Finally, the sixth and seventh columns contain the minimum and the maximum F1 score measured during the evaluation.

Table 1 shows the results obtained on in-domain models; while in Table 2, we present the results obtained by testing our approach on out-model domains.

In Table 2, the results about the DBP baseline are not reported because the used test set contains texts belonging to domains that are not included in the model.

By considering the overall results obtained on the IMD (Table 1), we may observe how the proposed approach outperforms the provided baselines. The measured F1 scores (average, minimum, and maximum) are higher of about 4% with respect to the baselines. The same happens for the Precision value, while for the Recall, all systems are very close to 100% and now significant differences have been detected. Test instances for which polarity has not been estimated

Table 1. Comparison between the results obtained by the three baselines and the ones obtained by the proposed system on in-model domains.

Approach	Avg. precision	Avg. recall	Avg. F1
MFP	0.5000	1.0000	0.6667
DBP	0.7218	0.9931	0.8352
DDP	0.7115	0.9946	0.8290
DAP	**0.7686**	**0.9984**	**0.8679**
Approach	Avg. deviation	Avg. min. F1	Avg. max. F1
MFP	0.0000	0.6667	0.6667
DBP	0.5028	0.8153	0.8543
DDP	0.5584	0.8121	0.8456
DAP	**0.5954**	**0.8469**	**0.8881**

Table 2. Comparison between the results obtained by the two baselines and the ones obtained by the proposed system on out-model domains.

Approach	Avg. precision	Avg. recall	Avg. F1
MFP	0.5000	1.0000	0.6667
DBP	–	–	–
DDP	0.6766	0.9931	0.8045
DAP	**0.7508**	**0.9985**	**0.8564**
Approach	Avg. deviation	Avg. min. F1	Avg. max. F1
MFP	0.0000	0.6667	0.6667
DDP	0.5796	0.7906	0.8198
DAP	**0.6060**	**0.8389**	**0.8755**

have been judged as "neutral". Concerning the stability of the approach, we can notice that the exploitation of the domain information leads to a lower deviation over the five folds.

The second overall evaluation concerns the analysis of the results obtained on the set of OMD. The interesting aspect of this evaluation is to measure how the system is able to address the task of detecting polarities of documents coming from a different set of domains with respect to the ones used to build the model. Results are shown in Table 2. The first thing that we may observe is how the effectiveness obtained by the proposed system is very close to the one obtained in the IMD evaluation. Indeed, the difference between the two F1 averages is only around 1%. This aspect remarked the capability of the proposed approach to work in a cross-domain environment and to exploit the linguistic overlaps between domains for estimating text polarities. In this second evaluation, it is also possible to notice how the DDP baseline obtained lower results with respect to the evaluation on the IMD. This result confirms that a solution based on exploiting information coming from a domain resulting the "most similar" to the text to analyze are inadequate for computing text polarities.

In light of these results, we may state that the exploitation of linguistic overlaps between domains is a suitable solution for compensating the possible lack of knowledge had by building opinion models on limited training sets.

5.4   The IRMUDOSA System at ESWC-2016 SSA Challenge Task #1

The system participated to the Task #1 of the Semantic Sentiment Analysis Challenge co-located with ESWC 2016. Table 3 shows the results of Task #1. We may observe that the system ranked third, not far from the best performer. This result confirm the viability of the implemented approach for future implementation after a study of the main error scenarios in which the fuzzy-based algorithm performed poorly.

Table 3. Precision-recall analysis and winners for Task 1.

System	Precision	Recall	F-measure
Efstratios Sygkounas, Xianglei Li Giuseppe Rizzo and Raphaël Troncy	0.85686	0.90541	0.88046
Emanuele Di Rosa and Alberto Durante	0.82777	0.90789	0.87142
Giulio Petrucci and Mauro Dragoni	**0.81837**	**0.89198**	**0.85359**
Andi Rexha, Mark Kröll Mauro Dragoni and Roman Kern	0.50494	0.81665	0.62403

6   Conclusion and Future Work

In this article, we presented an approach to multi-domain opinion mining exploiting linguistic overlaps between domains for estimating the polarity of texts. The approach is supported by the implementation of a fuzzy model used for representing either the polarity of each feature with respect to a particular domain and its associated uncertainty.

Models are built by combining information extracted from a training set with the knowledge contained in two supervised linguistic resources, Sentic.net and the General Inquirer. The estimation of polarities is performed by combining the degree which a text belongs to each domain with each domain-specific polarity information extracted from the model.

Results shown the effectiveness improvement of the proposed approach with respect to the baselines demonstrating its viability and the close gap between the proposed system and the best performer participated to the Task #1 of Semantic Sentiment Analysis Challenge proved the potential of the fuzzy-based solution. Moreover, the protocol used for the evaluation enables an easy reproducibility of the experiments and the comparison of obtained results with other systems.

Future work will focus either on the enrichment of the knowledge used for building the models and on the use of fuzzy membership functions. Finally, we foresee the integration of a concept extraction approach in order to equip the system with further semantic capabilities of extracting finer-grained information (i.e., single aspects and semantic information associated with them) which can be used during the model construction.

References

1. Pang, B., Lee, L., Vaithyanathan, S.: Thumbs up? Sentiment classification using machine learning techniques. In: Proceedings of EMNLP, Philadelphia, pp. 79–86. Association for Computational Linguistics, July 2002
2. Blitzer, J., Dredze, M., Pereira, F.: Biographies, bollywood, boom-boxes and blenders: domain adaptation for sentiment classification. In: ACL, pp. 187–205 (2007)
3. Pan, S.J., Ni, X., Sun, J.T., Yang, Q., Chen, Z.: Cross-domain sentiment classification via spectral feature alignment. In: WWW, pp. 751–760 (2010)
4. Liu, B., Zhang, L.: A survey of opinion mining and sentiment analysis. In: Aggarwal, C.C., Zhai, C.X. (eds.) Mining Text Data, pp. 415–463. Springer, Boston (2012). doi:10.1007/978-1-4614-3223-4_13
5. Pang, B., Lee, L.: A sentimental education: sentiment analysis using subjectivity summarization based on minimum cuts. In: ACL, pp. 271–278 (2004)
6. Dave, K., Lawrence, S., Pennock, D.M.: Mining the peanut gallery: opinion extraction and semantic classification of product reviews. In: WWW, pp. 519–528 (2003)
7. Paltoglou, G., Thelwall, M.: A study of information retrieval weighting schemes for sentiment analysis. In: ACL, pp. 1386–1395 (2010)
8. Tan, S., Wang, Y., Cheng, X.: Combining learn-based and lexicon-based techniques for sentiment detection without using labeled examples. In: SIGIR, pp. 743–744 (2008)
9. Qiu, L., Zhang, W., Hu, C., Zhao, K.: SELC: a self-supervised model for sentiment classification. In: CIKM, pp. 929–936 (2009)
10. Melville, P., Gryc, W., Lawrence, R.D.: Sentiment analysis of blogs by combining lexical knowledge with text classification. In: KDD, pp. 1275–1284 (2009)
11. Taboada, M., Brooke, J., Tofiloski, M., Voll, K.D., Stede, M.: Lexicon-based methods for sentiment analysis. Comput. Linguist. **37**(2), 267–307 (2011)
12. Somasundaran, S.: Discourse-level relations for Opinion Analysis. Ph.D. thesis, University of Pittsburgh (2010)
13. Wang, H., Zhou, G.: Topic-driven multi-document summarization. In: IALP, pp. 195–198 (2010)
14. Dragoni, M.: SHELLFBK: an information retrieval-based system for multi-domain sentiment analysis. In: Proceedings of the 9th International Workshop on Semantic Evaluation (SemEval 2015), Denver, pp. 502–509. Association for Computational Linguistics, June 2015

15. Petrucci, G., Dragoni, M.: An information retrieval-based system for multi-domain sentiment analysis. In: Gandon, F., Cabrio, E., Stankovic, M., Zimmermann, A. (eds.) SemWebEval 2015. CCIS, vol. 548, pp. 234–243. Springer, Cham (2015). doi:10.1007/978-3-319-25518-7_20

16. Rexha, A., Kröll, M., Dragoni, M., Kern, R.: Exploiting propositions for opinion mining. In: Sack, H., Dietze, S., Tordai, A., Lange, C. (eds.) SemWebEval 2016. CCIS, vol. 641, pp. 121–125. Springer, Cham (2016). doi:10.1007/978-3-319-46565-4_9

17. Federici, M., Dragoni, M.: A knowledge-based approach for aspect-based opinion mining. In: Sack, H., Dietze, S., Tordai, A., Lange, C. (eds.) SemWebEval 2016. CCIS, vol. 641, pp. 141–152. Springer, Cham (2016). doi:10.1007/978-3-319-46565-4_11

18. Dragoni, M., Tettamanzi, A.G., da Costa Pereira, C.: Propagating and aggregating fuzzy polarities for concept-level sentiment analysis. Cogn. Comput. **7**(2), 186–197 (2015)

19. Dragoni, M., Tettamanzi, A.G.B., da Costa Pereira, C.: A fuzzy system for concept-level sentiment analysis. In: Presutti, V., et al. (eds.) SemWebEval 2014. CCIS, vol. 475, pp. 21–27. Springer, Cham (2014). doi:10.1007/978-3-319-12024-9_2

20. Petrucci, G., Dragoni, M.: The IRMUDOSA system at ESWC-2016 challenge on semantic sentiment analysis. In: Sack, H., Dietze, S., Tordai, A., Lange, C. (eds.) SemWebEval 2016. CCIS, vol. 641, pp. 126–140. Springer, Cham (2016). doi:10.1007/978-3-319-46565-4_10

21. da Costa Pereira, C., Dragoni, M., Pasi, G.: A prioritized "And" aggregation operator for multidimensional relevance assessment. In: Serra, R., Cucchiara, R. (eds.) AI*IA 2009. LNCS, vol. 5883, pp. 72–81. Springer, Heidelberg (2009). doi:10.1007/978-3-642-10291-2_8

22. Federici, M., Dragoni, M.: Towards unsupervised approaches for aspects extraction. In: Dragoni, M., Recupero, D.R., Denecke, K., Deng, Y., Declerck, T. (eds.) Joint Proceedings of the 2nd Workshop on Emotions, Modality, Sentiment Analysis and the Semantic Web and the 1st International Workshop on Extraction and Processing of Rich Semantics from Medical Texts Co-located with ESWC 2016, Heraklion, 29 May 2016. CEUR Workshop Proceedings, vol. 1613. CEUR-WS.org (2016)

23. Federici, M., Dragoni, M.: A branching strategy for unsupervised aspect-based sentiment analysis. In: Dragoni, M., Recupero, D.R. (eds.) Proceedings of the 3rd International Workshop on Emotions, Modality, Sentiment Analysis and the Semantic Web Co-located with 14th ESWC 2017, Portroz, 28 May 2017. CEUR Workshop Proceedings, vol. 1874. CEUR-WS.org (2017)

24. Riloff, E., Patwardhan, S., Wiebe, J.: Feature subsumption for opinion analysis. In: EMNLP, pp. 440–448 (2006)

25. Wilson, T., Wiebe, J., Hwa, R.: Recognizing strong and weak opinion clauses. Comput. Intell. **22**(2), 73–99 (2006)

26. Palmero Aprosio, A., Corcoglioniti, F., Dragoni, M., Rospocher, M.: Supervised opinion frames detection with RAID. In: Gandon, F., Cabrio, E., Stankovic, M., Zimmermann, A. (eds.) SemWebEval 2015. CCIS, vol. 548, pp. 251–263. Springer, Cham (2015). doi:10.1007/978-3-319-25518-7_22

27. Hatzivassiloglou, V., Wiebe, J.: Effects of adjective orientation and gradability on sentence subjectivity. In: COLING, pp. 299–305 (2000)

28. Kim, S.M., Hovy, E.H.: Crystal: analyzing predictive opinions on the web. In: EMNLP-CoNLL, pp. 1056–1064 (2007)

29. Rexha, A., Kröll, M., Dragoni, M., Kern, R.: Polarity classification for target phrases in tweets: a Word2Vec approach. In: Sack, H., Rizzo, G., Steinmetz, N., Mladenić, D., Auer, S., Lange, C. (eds.) ESWC 2016. LNCS, vol. 9989, pp. 217–223. Springer, Cham (2016). doi:10.1007/978-3-319-47602-5_40

30. Rexha, A., Kröll, M., Kern, R., Dragoni, M.: An embedding approach for microblog polarity classification. In: Dragoni, M., Recupero, D.R. (eds.) Proceedings of the 3rd International Workshop on Emotions, Modality, Sentiment Analysis and the Semantic Web Co-located with 14th ESWC 2017, Portroz, 28 May 2017. CEUR Workshop Proceedings, vol. 1874. CEUR-WS.org (2017)

31. Dragoni, M., Reforgiato Recupero, D.: Challenge on fine-grained sentiment analysis within ESWC2016. In: Sack, H., Dietze, S., Tordai, A., Lange, C. (eds.) SemWebEval 2016. CCIS, vol. 641, pp. 79–94. Springer, Cham (2016). doi:10.1007/978-3-319-46565-4_6

32. Jakob, N., Gurevych, I.: Extracting opinion targets in a single and cross-domain setting with conditional random fields. In: EMNLP, pp. 1035–1045 (2010)

33. Jin, W., Ho, H.H., Srihari, R.K.: Opinionminer: a novel machine learning system for web opinion mining and extraction. In: KDD, pp. 1195–1204 (2009)

34. Liu, B., Hu, M., Cheng, J.: Opinion observer: analyzing and comparing opinions on the web. In: WWW, pp. 342–351 (2005)

35. Wu, Y., Zhang, Q., Huang, X., Wu, L.: Phrase dependency parsing for opinion mining. In: EMNLP, pp. 1533–1541 (2009)

36. Su, Q., Xu, X., Guo, H., Guo, Z., Wu, X., Zhang, X., Swen, B., Su, Z.: Hidden sentiment association in Chinese web opinion mining. In: WWW, pp. 959–968 (2008)

37. Dragoni, M., Azzini, A., Tettamanzi, A.G.B.: A novel similarity-based crossover for artificial neural network evolution. In: Schaefer, R., Cotta, C., Kołodziej, J., Rudolph, G. (eds.) PPSN 2010. LNCS, vol. 6238, pp. 344–353. Springer, Heidelberg (2010). doi:10.1007/978-3-642-15844-5_35

38. Qiu, G., Liu, B., Bu, J., Chen, C.: Opinion word expansion and target extraction through double propagation. Comput. Linguist. 37(1), 9–27 (2011)

39. Dragoni, M.: A three-phase approach for exploiting opinion mining in computational advertising. IEEE Intell. Syst. 32(3), 21–27 (2017)

40. Dragoni, M., Petrucci, G.: A neural word embeddings approach for multi-domain sentiment analysis. IEEE Trans. Affect. Comput. PP(99), 1 (2017)

41. Barbosa, L., Feng, J.: Robust sentiment detection on Twitter from biased and noisy data. In: COLING (Posters), pp. 36–44 (2010)

42. Bermingham, A., Smeaton, A.F.: Classifying sentiment in microblogs: is brevity an advantage? In: CIKM, pp. 1833–1836 (2010)

43. Go, A., Bhayani, R., Huang, L.: Twitter sentiment classification using distant supervision. CS224N Project Report, Standford University (2009)

44. Cambria, E., Hussain, A.: Sentic Computing: A Common-Sense-Based Framework for Concept-Level Sentiment Analysis. Springer, Cham (2015)

45. Cambria, E., Hussain, A.: Sentic album: content-, concept-, and context-based online personal photo management system. Cogn. Comput. 4(4), 477–496 (2012)

46. Wang, Q.F., Cambria, E., Liu, C.L., Hussain, A.: Common sense knowledge for handwritten Chinese recognition. Cogn. Comput. 5(2), 234–242 (2013)

47. Yoshida, Y., Hirao, T., Iwata, T., Nagata, M., Matsumoto, Y.: Transfer learning for multiple-domain sentiment analysis–identifying domain dependent/independent word polarity. In: AAAI, pp. 1286–1291 (2011)

48. Ponomareva, N., Thelwall, M.: Semi-supervised vs. cross-domain graphs for sentiment analysis. In: RANLP, pp. 571–578 (2013)

49. Huang, S., Niu, Z., Shi, C.: Automatic construction of domain-specific sentiment lexicon based on constrained label propagation. Knowl. Based Syst. **56**, 191–200 (2014)
50. Dragoni, M., da Costa Pereira, C., Tettamanzi, A.G.B., Villata, S.: Smack: an argumentation framework for opinion mining. In: Kambhampati, S. (ed.) Proceedings of the Twenty-Fifth International Joint Conference on Artificial Intelligence (IJCAI 2016), New York, 9–15 July 2016, pp. 4242–4243. IJCAI/AAAI Press (2016)
51. Cambria, E., Olsher, D., Rajagopal, D.: Senticnet 3: a common and common-sense knowledge base for cognition-driven sentiment analysis. In: AAAI, pp. 1515–1521 (2014)
52. Stone, P.J., Dunphy, D., Smith, M.: The General Inquirer: A Computer Approach to Content Analysis. M.I.T Press, Oxford (1966)
53. Zadeh, L.A.: Fuzzy sets. Inf. Control **8**, 338–353 (1965)
54. Manning, C.D., Surdeanu, M., Bauer, J., Finkel, J., Bethard, S.J., McClosky, D.: The Stanford CoreNLP natural language processing toolkit. In: Proceedings of 52nd Annual Meeting of the Association for Computational Linguistics: System Demonstrations, Baltimore, pp. 55–60. Association for Computational Linguistics, June 2014
55. van Rijsbergen, C.J.: Information Retrieval. Butterworth, London (1979)
56. Zadeh, L.A.: The concept of a linguistic variable and its application to approximate reasoning - I. Inf. Sci. **8**(3), 199–249 (1975)
57. Hellendoorn, H., Thomas, C.: Defuzzification in fuzzy controllers. Intell. Fuzzy Syst. **1**, 109–123 (1993)
58. Dragoni, M., Tettamanzi, A., da Costa Pereira, C.: Dranziera: an evaluation protocol for multi-domain opinion mining. In: Calzolari, N., Choukri, K., Declerck, T., Goggi, S., Grobelnik, M., Maegaard, B., Mariani, J., Mazo, H., Moreno, A., Odijk, J., Piperidis, S. (eds.) Proceedings of the Tenth International Conference on Language Resources and Evaluation (LREC 2016), Paris. European Language Resources Association (ELRA), May 2016

Opinion Mining with a Clause-Based Approach

Andi Rexha[1], Mark Kröll[1], Mauro Dragoni[2(✉)], and Roman Kern[1]

[1] Know-Center GmbH, Graz, Austria
{arexha,mkroell,rkern}@know-center.at
[2] FBK-IRST, Trento, Italy
dragoni@fbk.eu

Abstract. With different social media and commercial platforms, users express their opinion about products in a textual form. Automatically extracting the polarity (i.e. whether the opinion is positive or negative) of a user can be useful for both actors: the online platform incorporating the feedback to improve their product as well as the client who might get recommendations according to his or her preferences. Different approaches for tackling the problem, have been suggested mainly using syntactic features. The "Challenge on Semantic Sentiment Analysis" aims to go beyond the word-level analysis by using semantic information. In this paper we propose a novel approach by employing the semantic information of grammatical unit called preposition. We try to derive the *target* of the review from the *summary information*, which serves as an input to identify the proposition in it. Our implementation relies on the hypothesis that the proposition expressing the target of the summary, usually containing the main polarity information.

1 Introduction

User's opinions can be found in various social media platforms and online stores in textual form. The length and style of the text can vary substantially, ranging from short twitter messages to longer book reviews. They also refer to different aspects, from politics, to pictures and commercial products. The nature of these opinions change the way to analyze the text from automatic polarity detection systems. Twitter messages aren't expressed using syntactically correct text and require a different preprocessing than, for example, book reviews. For the "Challenge on Semantic Sentiment Analysis" the task (with the winners of recent years [1,7,18,46]) is to detect the polarity of user's opinions of products on Amazon.com reviews. The dataset [17] consists of a set of summaries about different topics. The review is compound by an *id*, a *summary* and a *textual description* and is represented in a XML format as shown in the Example 1.

Since the summary text is expressed in well formed text, Natural Language Processing (NLP) tools can be used to preprocess and analyze those reviews. For this challenge we use a two step approach. In the first step we isolate the syntactic information (proposition) in which the summary is expressed. In the second step we use a supervised approach in order to classify the reviews in positive or negative polarity.

© Springer International Publishing AG 2017
M. Dragoni et al. (Eds.): SemWebEval 2017, CCIS 769, pp. 166–175, 2017.
https://doi.org/10.1007/978-3-319-69146-6_15

```
<summary>Transformers</summary>
<text>By most accounts, the Michael Bay-directed Transformers
    films to date films to date are not very good, but that
    hasnt stopped them from making gobs and gobs of cash.
</text>
<polarity>positive</polarity>
```

Example 1. Example of a single entry in the dataset provided by the challange

The paper is organized in four sections. In Sect. 3 we detail the approach used for the two steps and the features used for the supervised task. Finally, Sect. 4 describes the partial results and the discussion the advantages and drawback of the approach.

2   Related Work

The topic of sentiment analysis has been studied extensively in the literature [31], where several techniques have been proposed and validated.

Machine learning techniques are the most common approaches used for addressing this problem, given that any existing supervised methods can be applied to sentiment classification. For instance, in [36], the authors compared the performance of Naive-Bayes, Maximum Entropy, and Support Vector Machines in sentiment analysis on different features like considering only unigrams, bigrams, combination of both, incorporating parts of speech and position information or by taking only adjectives. Moreover, beside the use of standard machine learning method, researchers have also proposed several custom techniques specifically for sentiment classification, like the use of adapted score function based on the evaluation of positive or negative words in product reviews [9], as well as by defining weighting schemata for enhancing classification accuracy [34].

An obstacle to research in this direction is the need of labeled training data, whose preparation is a time-consuming activity. Therefore, in order to reduce the labeling effort, opinion words have been used for training procedures. In [41,50], the authors used opinion words to label portions of informative examples for training the classifiers. Opinion words have been exploited also for improving the accuracy of sentiment classification, as presented in [33], where a framework incorporating lexical knowledge in supervised learning to enhance accuracy has been proposed. Opinion words have been used also for unsupervised learning approaches like the one presented in [49].

Another research direction concerns the exploitation of discourse-analysis techniques. [47] discusses some discourse-based supervised and unsupervised approaches for opinion analysis; while in [51], the authors present an approach to identify discourse relations.

The approaches presented above are applied at the document-level [11,21,37, 42], i.e., the polarity value is assigned to the entire document content. However, in some case, for improving the accuracy of the sentiment classification, a more fine-grained analysis of a document is needed. Hence, the sentiment classification of the single sentences, has to be performed. In the literature, we may find approaches ranging from the use of fuzzy logic [18,19,38] to the use of aggregation techniques [8] for computing the score aggregation of opinion words. In the case of sentence-level sentiment classification, two different sub-tasks have to be addressed: (i) to determine if the sentence is subjective or objective, and (ii) in the case that the sentence is subjective, to determine if the opinion expressed in the sentence is positive, negative, or neutral. The task of classifying a sentence as subjective or objective, called "subjectivity classification", has been widely discussed in the literature [22,23,45,53] and systems implementing the capabilities of identifying opinion's holder, target, and polarity have been presented [1]. Once subjective sentences are identified, the same methods as for sentiment classification may be applied. For example, in [25] the authors consider gradable adjectives for sentiment spotting; while in [29,43,44] the authors built models to identify some specific types of opinions.

In the last years, with the growth of product reviews, the use of sentiment analysis techniques was the perfect floor for validating them in marketing activities [16]. However, the issue of improving the ability of detecting the different opinions concerning the same product expressed in the same review became a challenging problem. Such a task has been faced by introducing "aspect" extraction approaches that were able to extract, from each sentence, which is the aspect the opinion refers to. In the literature, many approaches have been proposed: conditional random fields (CRF) [27], hidden Markov models (HMM) [28], sequential rule mining [30], dependency tree kernels [54], clustering [48], and genetic algorithms [13]. In [12,40], two methods were proposed to extract both opinion words and aspects simultaneously by exploiting some syntactic relations of opinion words and aspects.

A particular attention should be given also to the application of sentiment analysis in social networks [12,15]. More and more often, people use social networks for expressing their moods concerning their last purchase or, in general, about new products. Such a social network environment opened up new challenges due to the different ways people express their opinions, as described by [2,3], who mention "noisy data" as one of the biggest hurdles in analyzing social network texts.

One of the first studies on sentiment analysis on micro-blogging websites has been discussed in [24], where the authors present a distant supervision-based approach for sentiment classification.

At the same time, the social dimension of the Web opens up the opportunity to combine computer science and social sciences to better recognize, interpret, and process opinions and sentiments expressed over it. Such multi-disciplinary approach has been called *sentic computing* [6]. Application domains where sentic computing has already shown its potential are the cognitive-inspired classification of images [5], of texts in natural language, and of handwritten text [52].

Finally, an interesting recent research direction is domain adaptation, as it has been shown that sentiment classification is highly sensitive to the domain from which the training data is extracted. The reason is that words and even language constructs used in different domains for expressing opinions can be quite different. To make matters worse, the same word in one domain may have positive connotations, but in another domain may have negative ones; therefore, domain adaptation is needed. In the literature, different approaches related to the Multi-Domain sentiment analysis have been proposed. Briefly, two main categories may be identified: (i) the transfer of learned classifiers across different domains [4, 35, 55], and (ii) the use of propagation of labels through graph structures [14, 19, 26, 39].

All approaches presented above are based on the use of statistical techniques for building sentiment models. The exploitation of semantic information is not taken into account. In this work, we proposed a first version of a semantic-based approach preserving the semantic relationships between the terms of each sentence in order to exploit them either for building the model and for estimating document polarity. The proposed approach, falling into the multi-domain sentiment analysis category, instead of using pre-determined polarity information associated with terms, it learns them directly from domain-specific documents. Such documents are used for training the models used by the system.

3   Approach and Features

Each review is composed of a *summary* and a *textual information*. One or more sentences form the textual information of the summary contain the detailed specification of the user experience. We base our approach in the hypothesis that the summary is extended in the textual information and its "isolated" content contains the main polarity information. After preprocessing the textual summary with NLP tools, we annotate the words of the summary in each sentence. Later we extract the most "prominent" sentence (to be defined in Subsect. 3.2) which contains the main target of the summary. From the "prominent" sentence we select the "best fitting" proposition (we define it in Subsect. 3.3) which contains the summary. From the proposition, we extract the polarity of each word and encode the distribution of the polarity in the proposition as features. As a final step we train a classifier in order to predict the polarity of the whole tweet. Recapping, the approach can be split in the following steps:

- Preprocessing
- Extract the prominent sentence
- Extract the prominent proposition
- Polarity extraction and feature encoding

Below, we describe each of these steps in more detail. For illustration purposes we get the following example:

Summary: Typical movie of Al Pacino
Text: This was a very good movie from Al Pacino but the music wasn't that nice. Just think about how bad other movies are! The music doesn't play any role for my review!

Example 2. Example of a summary and review

3.1   Preprocessing

For preprocessing the reviews, we select the Stanford Core NLP tool [32]. For each review we annotate all sentences, words and parse the syntactic dependency graph. As a final step we annotate the text in the review with the tokens from the summary.

In the Example 2 this would be: This was a very good movie from Al Pacino but the music wasn't that nice.

3.2   Extract the Prominent Sentence

From the annotated text of the review we need to select the sentence best matching with the summary. We define the most "prominent" one as the sentence which contains more terms in a TermFrequency-InverseSentenceFrequency of the term. So, for each term of the summary we calculate it's frequency (i.e. the number of times it occurs in each sentence) and it's inverse sentence frequency (i.e. the inverse fraction of documents containing the word). This formula reflects the tf-idf (term frequency-inverse document frequency) score, but applied to the sentences, and we consider it a tf-isf. In the former example, the first sentence would be selected due to it containing two annotated words.

3.3   Extract the Prominent Proposition

For the "best fit" sentence we try to extract the *proposition* which capture best the main information. As a first step, we extract the propositions composing the sentence. We use the well known Open Information Extraction tool, ClausIE [10]. It extracts relation of the form (subject, predicate, object) called propositions.

Returning to the example of the prominent sentence "This was a very good movie from Al Pacino but the music wasn't that nice.", it can be splitted in the following propositions:

– This was a very good movie
– This was a very good movie from Al Pacino
– the music wasn't that nice

We define the "best matching" proposition as the shortest one (in terms of words) containing most of the terms in the summary. This mean that the selected one in our example would be: This was a very good movie from Al Pacino.

3.4 Features

In this challenge we use polarity features extracted from SentiWordNet [20]. SentiWordNet is a thesaurus which contains polarity information about words. To each word it is assigned a score between -1 and 1, which indicates whether the word has a negative or positive polarity. We model the proposition as a function of the sentiment expressed in the words. More precisely we identify the polarity of each word in the "best fit" proposition. We express the features of the "best fit" proposition as maximum, minimum, arithmetic mean, and standard deviation of the polarities of the words. As a additional feature we use the number of negation words expressed in the "best matching" sentence.

4 Results and Discussion

We try to learn our model from the features we have extracted and predict new unseen reviews. We use a Logistic Regression to learn from the results. In the Table 1 we present the precision, recall and F1-measure of the 10 fold cross-validation.

Table 1. Results from a 10 fold cross validation in the training dataset

	Precision	Recall	F1-measure
Positive	0.629	0.823	0.713
Negative	0.744	0.515	0.608
Average	0.686	0.669	0.661

As we can see from the tables, the results from the cross validation are not as good as expected. After an analyses of the dataset we believe that this discrepancy of the results from the expectation might be caused by the false assumption that the summary of the review is also expressed in the text.

Acknowledgment. This work is funded by the KIRAS program of the Austrian Research Promotion Agency (FFG) (project number 840824). The Know-Center is funded within the Austrian COMET Program under the auspices of the Austrian Ministry of Transport, Innovation and Technology, the Austrian Ministry of Economics and Labour and by the State of Styria. COMET is managed by the Austrian Research Promotion Agency FFG.

References

1. Palmero Aprosio, A., Corcoglioniti, F., Dragoni, M., Rospocher, M.: Supervised opinion frames detection with RAID. In: Gandon, F., Cabrio, E., Stankovic, M., Zimmermann, A. (eds.) SemWebEval 2015. CCIS, vol. 548, pp. 251–263. Springer, Cham (2015). doi:10.1007/978-3-319-25518-7_22

2. Barbosa, L., Feng, J.: Robust sentiment detection on twitter from biased and noisy data. In: COLING (Posters), pp. 36–44 (2010)
3. Bermingham, A., Smeaton, A.F.: Classifying sentiment in microblogs: is brevity an advantage? In: CIKM, pp. 1833–1836 (2010)
4. Blitzer, J., Dredze, M., Pereira, F.: Biographies, bollywood, boom-boxes and blenders: domain adaptation for sentiment classification. In: ACL, pp. 187–205 (2007)
5. Cambria, E., Hussain, A.: Sentic album: content-, concept-, and context-based online personal photo management system. Cogn. Comput. 4(4), 477–496 (2012)
6. Cambria, E., Hussain, A.: Sentic computing: a common-sense-based framework for concept-level sentiment analysis (2015)
7. Chung, J.K.-C., Wu, C.-E., Tsai, R.T.-H.: Polarity detection of online reviews using sentiment concepts: NCU IISR team at ESWC-14 challenge on concept-level sentiment analysis. In: Presutti, V., et al. (eds.) SemWebEval 2014. CCIS, vol. 475, pp. 53–58. Springer, Cham (2014). doi:10.1007/978-3-319-12024-9_7
8. da Costa Pereira, C., Dragoni, M., Pasi, G.: A prioritized "And" aggregation operator for multidimensional relevance assessment. In: Serra, R., Cucchiara, R. (eds.) AI*IA 2009. LNCS, vol. 5883, pp. 72–81. Springer, Heidelberg (2009). doi:10.1007/978-3-642-10291-2_8
9. Dave, K., Lawrence, S., Pennock, D.M.: Mining the peanut gallery: opinion extraction and semantic classification of product reviews. In: WWW, pp. 519–528 (2003)
10. Del Corro, L., Gemulla, R.: Clausie: clause-based open information extraction. In: Proceedings of the 22nd International Conference on World Wide Web (WWW 2013), pp. 355–366. ACM, New York (2013). http://doi.acm.org/10.1145/2488388.2488420
11. Dragoni, M.: SHELLFBK: an information retrieval-based system for multi-domain sentiment analysis. In: Proceedings of the 9th International Workshop on Semantic Evaluation (SemEval 2015), pp. 502–509. Association for Computational Linguistics, Denver, June 2015
12. Dragoni, M.: A three-phase approach for exploiting opinion mining in computational advertising. IEEE Intell. Syst. 32(3), 21–27 (2017). https://doi.org/10.1109/MIS.2017.46
13. Dragoni, M., Azzini, A., Tettamanzi, A.G.B.: A novel similarity-based crossover for artificial neural network evolution. In: Schaefer, R., Cotta, C., Kołodziej, J., Rudolph, G. (eds.) PPSN 2010. LNCS, vol. 6238, pp. 344–353. Springer, Heidelberg (2010). doi:10.1007/978-3-642-15844-5_35
14. Dragoni, M., da Costa Pereira, C., Tettamanzi, A.G.B., Villata, S.: SMACk: an argumentation framework for opinion mining. In: Kambhampati, S. (ed.) Proceedings of the Twenty-Fifth International Joint Conference on Artificial Intelligence (IJCAI 2016), New York, 9–15 July 2016, pp. 4242–4243. IJCAI/AAAI Press (2016). http://www.ijcai.org/Abstract/16/641
15. Dragoni, M., Petrucci, G.: A neural word embeddings approach for multi-domain sentiment analysis. IEEE Trans. Affect. Comput. PP(99), 1 (2017)
16. Dragoni, M., Reforgiato Recupero, D.: Challenge on fine-grained sentiment analysis within ESWC2016. In: Sack, H., Dietze, S., Tordai, A., Lange, C. (eds.) SemWebEval 2016. CCIS, vol. 641, pp. 79–94. Springer, Cham (2016). doi:10.1007/978-3-319-46565-4_6

17. Dragoni, M., Tettamanzi, A., da Costa Pereira, C.: Dranziera: an evaluation protocol for multi-domain opinion mining. In: Calzolari, N., Choukri, K., Declerck, T., Goggi, S., Grobelnik, M., Maegaard, B., Mariani, J., Mazo, H., Moreno, A., Odijk, J., Piperidis, S. (eds.) Proceedings of the Tenth International Conference on Language Resources and Evaluation (LREC 2016), Paris. European Language Resources Association (ELRA), May 2016

18. Dragoni, M., Tettamanzi, A.G.B., da Costa Pereira, C.: A fuzzy system for concept-level sentiment analysis. In: Presutti, V., et al. (eds.) SemWebEval 2014. CCIS, vol. 475, pp. 21–27. Springer, Cham (2014). doi:10.1007/978-3-319-12024-9_2

19. Dragoni, M., Tettamanzi, A.G., da Costa Pereira, C.: Propagating and aggregating fuzzy polarities for concept-level sentiment analysis. Cogn. Comput. **7**(2), 186–197 (2015). http://dx.doi.org/10.1007/s12559-014-9308-6

20. Esuli, A., Sebastiani, F.: SentiWordNet: a publicly available lexical resource for opinion mining. In: Proceedings of the 5th Conference on Language Resources and Evaluation (LREC 2006), pp. 417–422 (2006)

21. Federici, M., Dragoni, M.: A knowledge-based approach for aspect-based opinion mining. In: Sack, H., Dietze, S., Tordai, A., Lange, C. (eds.) SemWebEval 2016. CCIS, vol. 641, pp. 141–152. Springer, Cham (2016). doi:10.1007/978-3-319-46565-4_11

22. Federici, M., Dragoni, M.: Towards unsupervised approaches for aspects extraction. In: Dragoni, M., Recupero, D.R., Denecke, K., Deng, Y., Declerck, T. (eds.) Joint Proceedings of the 2th Workshop on Emotions, Modality, Sentiment Analysis and the Semantic Web and the 1st International Workshop on Extraction and Processing of Rich Semantics from Medical Texts Co-located with ESWC 2016, Heraklion, 29 May 2016. CEUR Workshop Proceedings, vol. 1613. CEUR-WS.org (2016). http://ceur-ws.org/Vol-1613/paper_2.pdf

23. Federici, M., Dragoni, M.: A branching strategy for unsupervised aspect-based sentiment analysis. In: Dragoni, M., Recupero, D.R. (eds.) Proceedings of the 3rd International Workshop on Emotions, Modality, Sentiment Analysis and the Semantic Web Co-located with 14th ESWC 2017, Portroz, 28 May 2017. CEUR Workshop Proceedings, vol. 1874. CEUR-WS.org (2017)

24. Go, A., Bhayani, R., Huang, L.: Twitter sentiment classification using distant supervision. CS224N Project Report, Standford University (2009)

25. Hatzivassiloglou, V., Wiebe, J.: Effects of adjective orientation and gradability on sentence subjectivity. In: COLING, pp. 299–305 (2000)

26. Huang, S., Niu, Z., Shi, C.: Automatic construction of domain-specific sentiment lexicon based on constrained label propagation. Knowl. Based Syst. **56**, 191–200 (2014)

27. Jakob, N., Gurevych, I.: Extracting opinion targets in a single and cross-domain setting with conditional random fields. In: EMNLP, pp. 1035–1045 (2010)

28. Jin, W., Ho, H.H., Srihari, R.K.: OpinionMiner: a novel machine learning system for web opinion mining and extraction. In: KDD, pp. 1195–1204 (2009)

29. Kim, S.M., Hovy, E.H.: Crystal: analyzing predictive opinions on the web. In: EMNLP-CoNLL, pp. 1056–1064 (2007)

30. Liu, B., Hu, M., Cheng, J.: Opinion observer: analyzing and comparing opinions on the web. In: WWW, pp. 342–351 (2005)

31. Liu, B., Zhang, L.: A survey of opinion mining and sentiment analysis. In: Aggarwal, C.C., Zhai, C.X. (eds.) Mining Text Data, pp. 415–463. Springer, Boston (2012). doi:10.1007/978-1-4614-3223-4_13

32. Manning, C.D., Surdeanu, M., Bauer, J., Finkel, J., Inc, P., Bethard, S.J., Mcclosky, D.: The Stanford coreNLP natural language processing toolkit. In: Proceedings of the 52nd Annual Meeting of the Association for Computational Linguistics: System Demonstrations, pp. 55–60 (2014)

33. Melville, P., Gryc, W., Lawrence, R.D.: Sentiment analysis of blogs by combining lexical knowledge with text classification. In: KDD, pp. 1275–1284 (2009)

34. Paltoglou, G., Thelwall, M.: A study of information retrieval weighting schemes for sentiment analysis. In: ACL, pp. 1386–1395 (2010)

35. Pan, S.J., Ni, X., Sun, J.T., Yang, Q., Chen, Z.: Cross-domain sentiment classification via spectral feature alignment. In: WWW, pp. 751–760 (2010)

36. Pang, B., Lee, L.: A sentimental education: sentiment analysis using subjectivity summarization based on minimum cuts. In: ACL, pp. 271–278 (2004)

37. Petrucci, G., Dragoni, M.: An information retrieval-based system for multi-domain sentiment analysis. In: Gandon, F., Cabrio, E., Stankovic, M., Zimmermann, A. (eds.) SemWebEval 2015. CCIS, vol. 548, pp. 234–243. Springer, Cham (2015). doi:10.1007/978-3-319-25518-7_20

38. Petrucci, G., Dragoni, M.: The IRMUDOSA system at ESWC-2016 challenge on semantic sentiment analysis. In: Sack, H., Dietze, S., Tordai, A., Lange, C. (eds.) SemWebEval 2016. CCIS, vol. 641, pp. 126–140. Springer, Cham (2016). doi:10. 1007/978-3-319-46565-4_10

39. Ponomareva, N., Thelwall, M.: Semi-supervised vs. cross-domain graphs for sentiment analysis. In: RANLP, pp. 571–578 (2013)

40. Qiu, G., Liu, B., Bu, J., Chen, C.: Opinion word expansion and target extraction through double propagation. Comput. Linguist. **37**(1), 9–27 (2011)

41. Qiu, L., Zhang, W., Hu, C., Zhao, K.: SELC: a self-supervised model for sentiment classification. In: CIKM, pp. 929–936 (2009)

42. Rexha, A., Kröll, M., Dragoni, M., Kern, R.: Exploiting propositions for opinion mining. In: Sack, H., Dietze, S., Tordai, A., Lange, C. (eds.) SemWebEval 2016. CCIS, vol. 641, pp. 121–125. Springer, Cham (2016). doi:10.1007/978-3-319-46565-4_9

43. Rexha, A., Kröll, M., Dragoni, M., Kern, R.: Polarity classification for target phrases in tweets: a Word2Vec approach. In: Sack, H., Rizzo, G., Steinmetz, N., Mladenić, D., Auer, S., Lange, C. (eds.) ESWC 2016. LNCS, vol. 9989, pp. 217–223. Springer, Cham (2016). doi:10.1007/978-3-319-47602-5_40

44. Rexha, A., Kröll, M., Kern, R., Dragoni, M.: An embedding approach for microblog polarity classification. In: Dragoni, M., Recupero, D.R. (eds.) Proceedings of the 3rd International Workshop on Emotions, Modality, Sentiment Analysis and the Semantic Web Co-located with 14th ESWC 2017, Portroz, 28 May 2017. CEUR Workshop Proceedings, vol. 1874. CEUR-WS.org (2017)

45. Riloff, E., Patwardhan, S., Wiebe, J.: Feature subsumption for opinion analysis. In: EMNLP, pp. 440–448 (2006)

46. Schouten, K., Frasincar, F.: The benefit of concept-based features for sentiment analysis. In: Gandon, F., Cabrio, E., Stankovic, M., Zimmermann, A. (eds.) SemWebEval 2015. CCIS, vol. 548, pp. 223–233. Springer, Cham (2015). doi:10. 1007/978-3-319-25518-7_19

47. Somasundaran, S.: Discourse-level relations for opinion analysis. Ph.D. thesis, University of Pittsburgh (2010)

48. Su, Q., Xu, X., Guo, H., Guo, Z., Wu, X., Zhang, X., Swen, B., Su, Z.: Hidden sentiment association in Chinese web opinion mining. In: WWW, pp. 959–968 (2008)

49. Taboada, M., Brooke, J., Tofiloski, M., Voll, K.D., Stede, M.: Lexicon-based methods for sentiment analysis. Comput. Linguist. **37**(2), 267–307 (2011)
50. Tan, S., Wang, Y., Cheng, X.: Combining learn-based and lexicon-based techniques for sentiment detection without using labeled examples. In: SIGIR, pp. 743–744 (2008)
51. Wang, H., Zhou, G.: Topic-driven multi-document summarization. In: IALP, pp. 195–198 (2010)
52. Wang, Q.F., Cambria, E., Liu, C.L., Hussain, A.: Common sense knowledge for handwritten Chinese recognition. Cogn. Comput. **5**(2), 234–242 (2013)
53. Wilson, T., Wiebe, J., Hwa, R.: Recognizing strong and weak opinion clauses. Comput. Intell. **22**(2), 73–99 (2006)
54. Wu, Y., Zhang, Q., Huang, X., Wu, L.: Phrase dependency parsing for opinion mining. In: EMNLP, pp. 1533–1541 (2009)
55. Yoshida, Y., Hirao, T., Iwata, T., Nagata, M., Matsumoto, Y.: Transfer learning for multiple-domain sentiment analysis–identifying domain dependent/independent word polarity. In: AAAI, pp. 1286–1291 (2011)

Author Index

Printed in the United States
By Bookmasters